Boolean Functions: Topics in Asynchronicity

Boolean Functions: Topics in Asynchronicity

First Edition

Serban E. Vlad
Oradea, Romania

WILEY

This edition first published 2019

Registered Office
John Wiley & Sons, Inc., 111 River Street, Hoboken, NJ 07030, USA

Editorial Office
111 River Street, Hoboken, NJ 07030, USA

For details of our global editorial offices, customer services, and more information about Wiley products visit us at www.wiley.com.

Wiley also publishes its books in a variety of electronic formats and by print-on-demand. Some content that appears in standard print versions of this book may not be available in other formats.

Library of Congress Cataloging-in-Publication Data

Names: Vlad, Serban E., 1959- author.
Title: Boolean functions : topics in asynchronicity / Serban E. Vlad.
Description: First edition. | Hoboken, NJ : John Wiley & Sons, 2018. |
Includes bibliographical references and index. |
Identifiers: LCCN 2018034871 (print) | LCCN 2018057135 (ebook) | ISBN 9781119517498 (Adobe PDF) | ISBN 9781119517511 (ePub) | ISBN 9781119517474 (hardcover) | ISBN 9781119517498 (ePDF)
Subjects: LCSH: Algebra, Boolean.
Classification: LCC QA10.3 (ebook) | LCC QA10.3 .V533 2018 (print) | DDC 511.3/24–dc23
LC record available at https://lccn.loc.gov/2018034871

ISBN 978-1-119-51747-4 (Hardback)
ISBN 978-1-119-51749-8 (ePDF)
ISBN 978-1-119-51751-1 (epub)

Cover Design: Wiley
Cover Image: Courtesy of Serban E. Vlad

Set in 10/12pt WarnockPro by SPi Global, Chennai, India

Printed in United States of America

V10007259_010419

To Ciupi and Puiu Mic

Contents

Preface

In this framework, by asynchronicity we mean that the coordinate functions $\Phi_1, \ldots, \Phi_n$ of a function Φ are computed independently on each other, asynchronously. Synchronicity is that special case of asynchronicity when $\Phi_1, \ldots, \Phi_n$ are computed at the same time, synchronously. It is of special interest to study the iterations of Φ, which can be asynchronous or synchronous.

Our project "Topics in asynchronicity" has been thought of having two parts, part I: Boolean functions and part II: Boolean systems. While working we took the decision to split it in two books.

The source of inspiration is represented by the asynchronous circuits from electronics that can be modeled by Boolean functions $\Phi : \{0,1\}^n \longrightarrow \{0,1\}^n$ iterating their coordinates in arbitrary time, independently on each other, i.e. asynchronously. The uncertainties related with the behavior of the circuits and their models are generated by technology and also by temperature variations and voltage supply variations.

In order to understand the dynamics of these systems, we give the example of the function Φ from Table 1, whose state portrait was drawn in Figure 1 (we have adopted the terminology of state portrait, by analogy with the phase portraits of the dynamical systems theory; such drawings might be called in engineering and elsewhere state transition graphs or state transition diagrams).

In Figure 1, the arrows show the increase of time. We have underlined in the tuples $(\mu_1, \mu_2, \mu_3) \in \{0,1\}^3$ these coordinates, called unstable (or excited, or enabled), for which $\mu_i \neq \Phi_i(\mu), i \in \{1,2,3\}$; these are the coordinates that are about to switch, but the time instant and the order in which these switches happen are not known. In this model, each present value $\mu \in \{0,1\}^3$ of the state may be followed by several possible values in the future, giving nondeterminism and also branching time in the future.

$(1,0,1)$ is an isolated fixed point of Φ (a fixed point is also called equilibrium point, or rest position, or final state), where the system stays indefinitely long; it has no underlined coordinates. Unlike it, $(0,1,0)$ is a fixed point that is not isolated, since a transfer to it exists, from $(0,0,0)$.

Table 1 An example.

(μ_1, μ_2, μ_3)	$\Phi(\mu_1, \mu_2, \mu_3)$
$(0, 0, 0)$	$(0, 1, 1)$
$(0, 0, 1)$	$(0, 1, 1)$
$(0, 1, 0)$	$(0, 1, 0)$
$(0, 1, 1)$	$(1, 1, 1)$
$(1, 0, 0)$	$(0, 0, 0)$
$(1, 0, 1)$	$(1, 0, 1)$
$(1, 1, 0)$	$(1, 0, 0)$
$(1, 1, 1)$	$(1, 1, 0)$

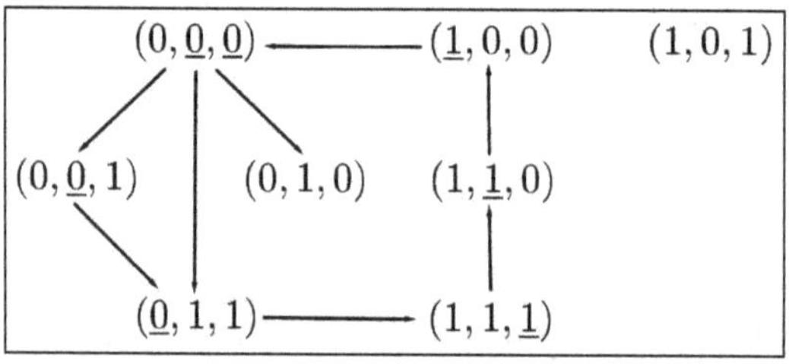

Figure 1 Dependence on the order in which Φ_1, Φ_2, Φ_3 are computed.

The transition $(0, 1, 1) \longrightarrow (1, 1, 1)$ consists in the computation of $\Phi_1(0, 1, 1)$; even if we do not know when it happens, we know that it happens and the system, if it is in $(0, 1, 1)$, surely gets to $(1, 1, 1)$ sometime. And the transitions $(1, 1, 1) \longrightarrow (1, 1, 0), (1, 1, 0) \longrightarrow (1, 0, 0), (1, 0, 0) \longrightarrow (0, 0, 0)$ are similar.

The interesting behavior is in $(0, 0, 0)$; if $\Phi_3(0, 0, 0)$ is computed first, or if $\Phi_2(0, 0, 0)$, $\Phi_3(0, 0, 0)$ are computed simultaneously, the system gets to $(0, 1, 1)$ sometime, with $(0, 0, 1)$ a possible intermediate state; but if $\Phi_2(0, 0, 0)$ is computed first, then the state $(0, 1, 0)$ is reached and, as it is a fixed point of Φ (no coordinate is underlined), the system rests there indefinitely long.

In the previous discussion:

(a) a system is identified with a function $\Phi : \{0, 1\}^n \longrightarrow \{0, 1\}^n$, in the sense that Φ contains all the information that gives the behavior of the system. This justifies our definitions of the Boolean functions via state portraits and, in fact, this is the motivation situated behind writing this book;

(b) the system that we refer to is Boolean (this vaguely refers at $\{0, 1\}$), universal (the state space is all of $\{0, 1\}^n$), regular (a generator function Φ exists), asynchronous ($\Phi_1, \dots, \Phi_n$ are not computed at the same time, but the fact that these coordinates are computed independently on each other also shows that the structure of the system is variable), nondeterministic ($\Phi_1, \dots, \Phi_n$ have unknown durations of computation), autonomous (no input), noninitialized;

(c) the durations of computation of $\Phi_1, .., \Phi_n$ are subject to no restriction (this is the unbounded delay model of computation of the Boolean functions);

(d) time is discrete or continuous.

The topics that are common for the Boolean functions and the Boolean systems include morphisms and antimorphisms, invariant sets, the conditions of proper operation (race-freedom) and time-reversal, with the symmetry that it generates.

The concept of isomorphism is easily understood by looking at Figure 2, where the state portrait of a function which is isomorphic with the function Φ from Figure 1 was drawn. To each point $\mu \in \{0,1\}^3$ from Figure 1, the vector $(1,0,1)$ was added modulo 2 coordinatewise and the result expressed by Figure 2 is that the transitions from the first case become transitions translated with $(1,0,1)$ in the second case. Note the arrows and the underlined coordinates of the two functions: the behavior is the same. The morphisms present under a more general form this transfer of properties from a function to another one and this is observed by taking a look at Figure 3, representing a function that accepts a morphism from it to both functions, from Figures 1 and 2. The morphism of this example forgets the computations of Φ_2.

A nonempty set $A \subset \{0,1\}^n$ is invariant (Φ is kept in mind) if, whenever a computation starts in $\mu \in A$, it ends in some point $\mu' \in A$ and two types of invariance are situated behind this intuition. We can think, looking at Figure 2, that the fixed points $(0,0,0), (1,1,1)$ give the invariant sets $\{(0,0,0)\}$, $\{(1,1,1)\}$, $\{(0,0,0),(1,1,1)\}$ (where the computation starts in $(0,0,0)$ and

Figure 2 This function is isomorphic with the function Φ from Figure 1.

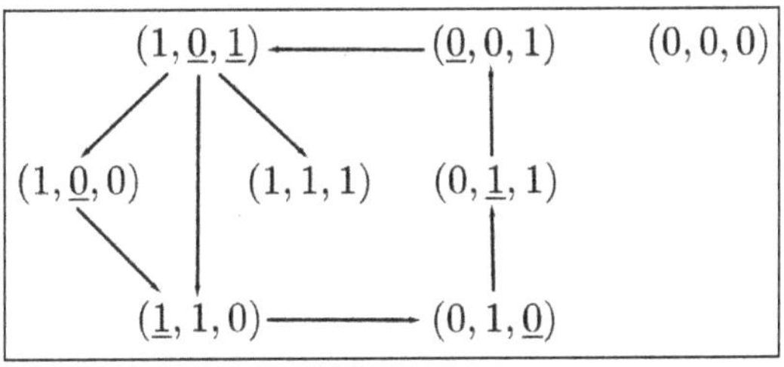

Figure 3 A morphism exists from this function to the functions from Figure 1 and Figure 2: the computation of Φ_2 is forgotten.

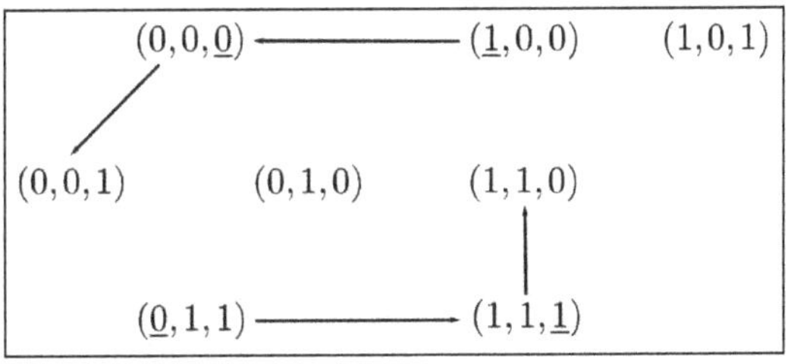

ends in (0, 0, 0) etc.). But the set $\{0,1\}^3\backslash\{(0,0,0)\}$ is also invariant. From the previous sets, $\{(0,0,0)\}$, $\{(1,1,1)\}$ are attractors.

The functions from Figures 1 and 2 suggest the problem of finding sets of Boolean functions Φ where, even if we do not know the time instants and the order in which their coordinates are computed, we know that for any $\mu \in \{0,1\}^n$, the values $\mu, \Phi(\mu), (\Phi \circ \Phi)(\mu), (\Phi \circ \Phi \circ \Phi)(\mu), \ldots$ are computed sometime, in this order. Thus, the behavior of the (asynchronous) systems that we are looking for reproduces in a certain way the behavior of the (synchronous, usual) dynamical systems in their Boolean version, and this is considered to be "nice", in a context with many unknown parameters. We get the "proper operation" properties of the Boolean functions/systems. The functions from Figures 1 and 2 do not fulfill such a property, for example in Figure 2 it is not sure that $\Phi(1,0,1) = (1,1,0)$ is really computed, since the computation of $\Phi_2(1,0,1)$ first produces the transfer of the system from (1, 0, 1) to (1, 1, 1), where it rests indefinitely long. The functions from Figures 3 and 4 fulfill the proper operation property.

Time reversal means, roughly speaking, reversing the arrows of a state portrait. The function from Figure 2 does not accept this, since reversing the arrows that point to (1, 1, 0) to arrows that start from (1, 1, 0) is impossible, but the function from Figure 4 does accept. We have inserted an arrow from (1, 1, 1) to (1, 1, 0). The time-reversed symmetrical function of the function from Figure 4 is the function from Figure 5.

The book is structured into chapters, sections, paragraphs, and the chapters have a short introduction and/or summary. The paragraphs are definitions,

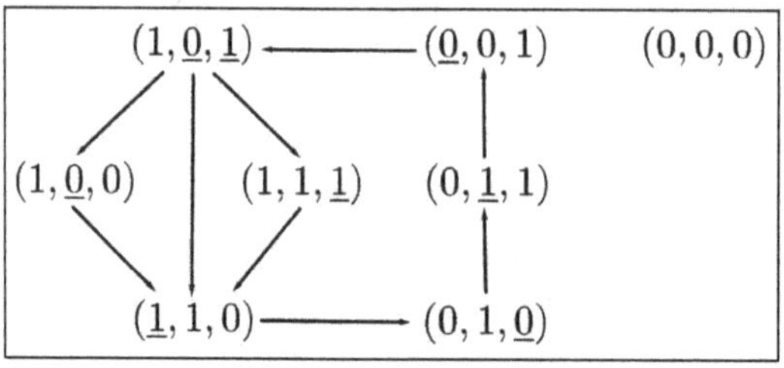

Figure 4 Function that accepts time reversal.

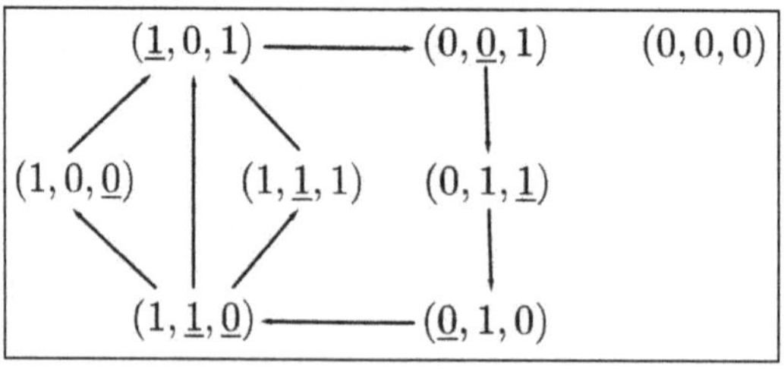

Figure 5 The time-reversed symmetrical function of the function from Figure 4.

notations, remarks and also theorems, lemmas, corollaries with their proofs. We have added two appendixes containing the definition of the category whose objects are $\Phi : \{0,1\}^n \longrightarrow \{0,1\}^n$ functions, and the notations respectively.

In Chapter 1, we introduce the binary Boole algebra and we give some preliminaries on Boolean functions (such as duality, iterates, state portraits, ...). We sketch how the asynchronous circuits are modeled by Boolean functions.

Chapter 2 is dedicated to the affine spaces $[\mu, \mu'] \subset \{0,1\}^n$ defined by two points $\mu, \mu' \in \{0,1\}^n$. Their meaning consists in the fact that, given $\Phi : \{0,1\}^n \longrightarrow \{0,1\}^n$, the set $[\mu, \Phi(\mu)]$ contains the points that may be accessed when $\Phi(\mu)$ is computed asynchronously; for example, looking at Figure 4, we have $[(0,0,0), \Phi(0,0,0)] = [(0,0,0),(0,0,0)] = \{(0,0,0)\}$ and also $[(1,0,1), \Phi(1,0,1)] = [(1,0,1),(1,1,0)] = \{(1,0,1),(1,0,0),(1,1,1),(1,1,0)\}$.

Chapters 3 and 4 introduce the morphisms and the antimorphisms of Boolean functions. Intuitively, the morphisms from Φ to Ψ presume that the succession of the cause $\mu, \nu \in \{0,1\}^n$ and the effect $\Phi(\mu), \Psi(\nu) \in \{0,1\}^n$ is the same, time flows in the same sense when Φ, Ψ are computed (from the past to the future, or from the future to the past); and the antimorphisms act as if for Φ and Ψ time flows in opposite senses.

The invariant sets are treated in Chapters 5 and 6. The concept is taken from the dynamical systems theory and it is brought in this timeless framework, where we discuss as well symmetry relative to translations, maximality and minimality, disconnectedness, etc. The path connected sets are treated in Chapter 7: the visual meaning of a path connecting $\mu \in \{0,1\}^n$ with $\nu \in \{0,1\}^n$ is given by the existence in the state portrait of the points $\lambda^1, \ldots, \lambda^p \in \{0,1\}^n$ such that an arrow exists from μ to λ^1 and ... and an arrow exists from λ^p to ν. Chapter 8 introduces the attractors, the nonempty subsets of $\{0,1\}^n$ satisfying a strong version of invariance together with one of: topological transitivity, minimality, and path connectedness.

Four concepts of proper operation of a Boolean function are presented in Chapters 9–12. Their fulfillment decreases the nondeterminism that is generated by the unknown parameters that occur in modeling.

Time-reversal symmetry is addressed in Chapter 13 and, combined with the fulfillment of the proper operation properties, in Chapters 14 and 15. This concept that is originating in physics and in the dynamical systems theory refers to two functions Φ, Ψ, which are iterated in time flowing in opposite senses. The result is that Φ, Ψ are somehow inverse to each other. We have kept the terminology of time-reversal and time-reversal symmetry even in this timeless approach since other symmetries of the Boolean functions exist also.

The most significant part of the bibliography consists in literature dedicated to dynamical systems theory and we translate concepts from synchronous real numbers systems to asynchronous Boolean systems by making analogies, and then from Boolean systems to Boolean functions. We indicate in this respect [2, 12, 15, 16, 19, 25], and [40].[1]

An important monograph in Boolean functions is [7].

Introducing the asynchronous Boolean systems is the purpose of [35], see also [30].

The technical condition of proper operation was known for many years by the theoreticians in switching circuits [24], perhaps with different names (race-freedom is one of them). We have also gathered useful intuition in this direction from many engineering sources such as [5, 6, 27].

An excellent survey in the time-reversal symmetry of the dynamical systems is [20].

The book is addressed to mathematicians and computer scientists who are interested in theory and applications of Boolean functions, dynamical systems, and circuits.

Kind thanks to the unknown reviewers and the publisher for their elegant cooperation.

Oradea, May, 2018 *Serban E. Vlad*

---

1 The monographs [16, 25], and [40] were suggested by a reviewer of the book. Other two monographs suggested by him are: Eric Goles, Servet Martínez, Neural and Automata Networks: Dynamical Behavior and Applications, Kluwer Academic Publishers, 1990 and Robert Francois, Discrete Iterations: A Metric Study, Springer Series in Computational Mathematics, 1986, but at the moment when this text was written, they were not available to the author.

1

Boolean Functions

The works of George Boole (1815–1864) [8] are contained in about 50 articles and a few other publications. Calculus of reasoning, which Boole was preoccupied with, found its way into his 1847 work, *The Mathematical Analysis of Logic*, that continued the ideas of Gottfried Wilhelm Leibniz (1646–1716) and pushed the idea that logic was a mathematical discipline, rather than philosophy. In 1854, Boole published *An Investigation into the Laws of Thought, on Which are Founded the Mathematical Theories of Logic and Probabilities*, which is perhaps his most important work. Boole approached logic in a new way, reducing it to simple algebra, incorporating logic into mathematics, and laying the foundations of the now famous binary approach. Logical expressions are now represented using a mathematical form called in his honor Boolean Algebra.

Boole's work on mathematical logic was criticized and/or ignored by his contemporaries, except for an American logician, Charles Sanders Peirce (1839–1914), who gave a speech at the American Academy of Arts and Sciences, describing Boole's ideas. Peirce spent more than 20 years working on these ideas and their applications in electronic circuitry.[1]

Unfortunately, Boolean algebra remained mostly unknown and unused for many years, until the 1940s, when a young student by the name of Claude Elwood Shannon picked up Boole's and Peirce's works and recognized their relevance to electronics design.

The purpose of this chapter is that of giving the first definitions and notations related with the Boolean algebra with two elements $\mathbf{B}$.

The functions $\Phi : \mathbf{B}^n \longrightarrow \mathbf{B}^p$ are called Boolean, and $0, 1; \cdot, \cup$ are dual couples, while the logical complement — is self-dual.

The iterates of $\Phi : \mathbf{B}^n \longrightarrow \mathbf{B}^n$ are two kinds: the first given by the composition $\Phi \circ \Phi$ and the second by allowing some coordinates Φ_i to be computed and some others Φ_j to be not.

The state portrait of $\Phi : \mathbf{B}^n \longrightarrow \mathbf{B}^n$ is a directed graph that indicates all the associations resulting when $1, \ldots, n$ coordinates of Φ are computed. The graph

1 Peirce is considered also the "father of pragmatism" and one of the founders of semiotics.

Boolean Functions: Topics in Asynchronicity, First Edition. Serban E. Vlad.

is associated with a unique Φ, thus it may act as a definition of Φ. The state portraits are widely used to characterize the asynchronous computations.

An example is given of the way that the asynchronous circuits are modeled by the Boolean functions.

The predecessors and successors, as well as the sources, isolated fixed points, transient points and sinks are defined. These concepts characterize the "past" possible causes and the "future" possible effects around a point μ.

In the last section, we introduce the translations.

1.1 The Binary Boole Algebra

Notation 1.1 We denote $\mathbf{B} = \{0, 1\}$.

Definition 1.1 The set **B** and, more general, the set $\mathbf{B}^n, n \geq 1$ are organized as topological spaces by the discrete topology.

Definition 1.2 The following laws are defined on **B**:

$$\begin{array}{cc} - & \\ 0 & 1 \\ 1 & 0 \end{array} ; \quad \begin{array}{ccc} \cdot & 0 & 1 \\ 0 & 0 & 0 \\ 1 & 0 & 1 \end{array} ; \quad \begin{array}{ccc} \cup & 0 & 1 \\ 0 & 0 & 1 \\ 1 & 1 & 1 \end{array} ; \quad \begin{array}{ccc} \oplus & 0 & 1 \\ 0 & 0 & 1 \\ 1 & 1 & 0 \end{array}$$

Table 1.1

called **negation, not**, or **(logical) complement; product** or **intersection; sum** or **union**; and **modulo 2 sum** or **disjoint union**. These laws induce laws that are denoted with the same symbols on $\mathbf{B}^n$.

Remark 1.1 The laws of **B** fulfill well-known equations, such as $\forall \xi \in \mathbf{B}$, $\forall \xi' \in \mathbf{B}$,

$$\begin{aligned} \overline{\overline{\xi}} &= \xi, \\ \xi \oplus \xi' &= \overline{\xi}\xi' \cup \xi\overline{\xi'}, \\ \overline{\xi} &= \xi \oplus 1, \end{aligned}$$

etc.

Definition 1.3 Let I an arbitrary, finite set. We consider the (binary) family $a_i \in \mathbf{B}, i \in I$, which is denoted sometimes with a or with (a_i) when I is kept in mind.[2] The set

$$supp\ a = \{i | i \in I, a_i = 1\}$$

---

2 The family $a_i \in \mathbf{B}, i \in I$ is a function $a : I \longrightarrow \mathbf{B}$ with $\forall i \in I, a_i = a(i)$ and similarly for $a_i \in \mathbf{B}^n, i \in I$.

is called the **support** (**set**) of a. More general, the support of the family $a_i \in \mathbf{B}^n, i \in I$ is

$$supp\ a = \{i | i \in I, a_i \neq (0, \dots, 0)\}.$$

By definition, the support of $a_i, i \in \emptyset$ is

$$supp\ a = \emptyset.$$

Definition 1.4 The **modulo 2 summation** of the family $a_i \in \mathbf{B}, i \in I$, denoted with $\underset{i \in I}{\Xi} a_i$, is:

$$\underset{i \in I}{\Xi} a_i = \begin{cases} 1, \text{if } card(supp\ a) \text{ is odd,} \\ 0, \text{if } card(supp\ a) \text{ is even.} \end{cases}$$

By definition

$$\underset{i \in \emptyset}{\Xi} a_i = 0.$$

This summation induces a summation $\underset{i \in I}{\Xi} a_i$ of the families $a_i \in \mathbf{B}^n, i \in I$ where *supp a* is a finite set.

Remark 1.2 We have the following conventions, which are compatible with Definition 1.4 and which will be used throughout this work:

- the empty set $\emptyset$ is a finite set;
- 0 $(= card(\emptyset))$ is an even number.

Remark 1.3 As previously I is a finite set, $supp\ a \subset I$ is also finite and $card(supp\ a) \in \{0, 1, 2, \dots\}$, thus Definition 1.4 makes sense.

Notation 1.2 We denote $\varepsilon^i \in \mathbf{B}^n, i \in \{1, \dots, n\}$,

$$\varepsilon^i = (0, \dots, \underset{i}{1}, \dots, 0).$$

Remark 1.4 The set $\mathbf{B}$ is a Boole algebra together with the laws $\overline{}$, $\cdot$, $\cup$ and also a field together with $\oplus, \cdot$. On the other hand $\mathbf{B}^n$ is an n-dimensional linear space together with the laws $\mathbf{B}^n \times \mathbf{B}^n \ni (\mu, \nu) \mapsto \mu \oplus \nu \in \mathbf{B}^n$ and $\mathbf{B} \times \mathbf{B}^n \ni (\zeta, \mu) \mapsto \zeta \cdot \mu \in \mathbf{B}^n$,

$$\mu \oplus \nu = (\mu_1 \oplus \nu_1, \dots, \mu_n \oplus \nu_n), \tag{1.1}$$

$$\zeta \cdot \mu = (\zeta \cdot \mu_1, \dots, \zeta \cdot \mu_n). \tag{1.2}$$

The canonical basis of $\mathbf{B}^n$ is given by $\varepsilon^1, \dots, \varepsilon^n$.

Definition 1.5 The elements $\mu \in \mathbf{B}^n$ are called **points, vectors, tuples**, or n-**tuples**.

Notation 1.3 Given the points $\mu \in \mathbf{B}^n$ and $\lambda \in \mathbf{B}^n$, we use the notation

$$\mu \boxplus \lambda = \{i | i \in \{1, \ldots, n\}, \mu_i \neq \lambda_i\}.$$

Theorem 1.1 We have the properties:

$$\mu \boxplus \lambda = \lambda \boxplus \mu, \tag{1.3}$$

$$\mu \boxplus \mu = \emptyset, \tag{1.4}$$

$$\mu \boxplus \overline{\mu} = \{1, \ldots, n\}, \tag{1.5}$$

where the negation acts coordinatewise: $\overline{\mu} = (\overline{\mu_1}, \ldots, \overline{\mu_n})$, $\forall \tau \in \mathbf{B}^n$,

$$\mu \boxplus \lambda = (\tau \oplus \mu) \boxplus (\tau \oplus \lambda) \tag{1.6}$$

and also

$$\mu \oplus \lambda = \underset{i \in \mu \boxplus \lambda}{\Xi} \varepsilon^i. \tag{1.7}$$

Proof: For example, (1.5) is a consequence of $\mu \boxplus \overline{\mu} = \{i | i \in \{1, \ldots, n\}, \mu_i \neq \overline{\mu_i}\} = \{1, \ldots, n\}$. We prove (1.7):

$$\mu \oplus \lambda = \underset{i \in \{1,..,n\}}{\Xi} \mu_i \varepsilon^i \oplus \underset{i \in \{1,..,n\}}{\Xi} \lambda_i \varepsilon^i = \underset{i \in \{1,..,n\}}{\Xi} (\mu_i \oplus \lambda_i) \varepsilon^i$$

$$= \underset{i \in \{j | j \in \{1,..,n\}, \mu_j \oplus \lambda_j = 1\}}{\Xi} \varepsilon^i = \underset{i \in \mu \boxplus \lambda}{\Xi} \varepsilon^i$$

□

1.2 Definition of the Boolean Functions. Examples. Duality

Definition 1.6 The functions $\Phi : \mathbf{B}^n \longrightarrow \mathbf{B}^p$ are called **Boolean functions.**

Remark 1.5 When a function Φ is given, we may specify the notations of the variables that occur during a reasoning under the form of the association $\mathbf{B}^n \ni \mu \longmapsto \Phi(\mu) \in \mathbf{B}^p$. We have already done so at Remark 1.4.

Example 1.1 The identity function $1_{\mathbf{B}^n} : \mathbf{B}^n \longrightarrow \mathbf{B}^n$ is defined in the following way:

$$\forall \mu \in \mathbf{B}^n, 1_{\mathbf{B}^n}(\mu) = \mu. \tag{1.8}$$

It has the remarkable property that its set of fixed points coincides with $\mathbf{B}^n$.

Example 1.2 We take a point $\mu' \in \mathbf{B}^p$. The constant function $\Phi : \mathbf{B}^n \longrightarrow \mathbf{B}^p$ is defined by

$$\forall \mu \in \mathbf{B}^n, \Phi(\mu) = \mu'. \tag{1.9}$$

Example 1.3 The projection of $\mathbf{B}^n$ on the ith coordinate is the function $\pi_i : \mathbf{B}^n \to \mathbf{B}, i \in \{1, \ldots, n\}$ defined by

$$\forall \mu \in \mathbf{B}^n, \pi_i(\mu) = \mu_i. \tag{1.10}$$

Definition 1.7 The **dual** $\Phi^* : \mathbf{B}^n \longrightarrow \mathbf{B}^p$ of the function $\Phi : \mathbf{B}^n \longrightarrow \mathbf{B}^p$ is by definition

$$\forall \mu \in \mathbf{B}^n, \Phi^*(\mu) = \overline{\Phi(\overline{\mu})}, \tag{1.11}$$

in other words $\forall i \in \{1, \ldots, n\}, \Phi_i^*(\mu_1, \ldots, \mu_n) = \overline{\Phi_i(\overline{\mu_1}, \ldots, \overline{\mu_n})}$.

Remark 1.6 For any $\mu \in \mathbf{B}^n$,

$$(\Phi^*)^*(\mu) = \overline{\Phi^*(\overline{\mu})} = \overline{\overline{\Phi(\overline{\overline{\mu}})}} = \Phi(\mu).$$

Example 1.4 We give some examples of dual functions: $(1_{\mathbf{B}^n})^* = 1_{\mathbf{B}^n}$, the dual of the constant function equal with μ' is the constant function equal with $\overline{\mu'}$ and $(\pi_i)^* = \pi_i, i \in \{1, \ldots, n\}$. In addition, given $\Upsilon : \mathbf{B} \longrightarrow \mathbf{B}$ and $\Phi, \Omega, \Gamma : \mathbf{B}^2 \longrightarrow \mathbf{B}$,

$$\forall \mu \in \mathbf{B}, \Upsilon(\mu) = \overline{\mu},$$
$$\forall \mu \in \mathbf{B}^2, \Phi(\mu_1, \mu_2) = \mu_1 \mu_2,$$
$$\forall \mu \in \mathbf{B}^2, \Omega(\mu_1, \mu_2) = \mu_1 \cup \mu_2,$$
$$\forall \mu \in \mathbf{B}^2, \Gamma(\mu_1, \mu_2) = \mu_1 \oplus \mu_2,$$

we have

$$\forall \mu \in \mathbf{B}, \Upsilon^*(\mu) = \overline{\overline{\overline{\mu}}} = \overline{\mu},$$
$$\forall \mu \in \mathbf{B}^2, \Phi^*(\mu_1, \mu_2) = \overline{\Phi(\overline{\mu_1}, \overline{\mu_2})} = \overline{\overline{\mu_1}\ \overline{\mu_2}} = \overline{\overline{\mu_1}} \cup \overline{\overline{\mu_2}} = \mu_1 \cup \mu_2,$$
$$\forall \mu \in \mathbf{B}^2, \Omega^*(\mu_1, \mu_2) = \overline{\Omega(\overline{\mu_1}, \overline{\mu_2})} = \overline{\overline{\mu_1} \cup \overline{\mu_2}} = \overline{\overline{\mu_1}}\ \overline{\overline{\mu_2}} = \mu_1 \mu_2,$$
$$\begin{aligned}\forall \mu \in \mathbf{B}^2, \Gamma^*(\mu_1, \mu_2) &= \overline{\Gamma(\overline{\mu_1}, \overline{\mu_2})} \\ &= \overline{\overline{\mu_1} \oplus \overline{\mu_2}} = ((\mu_1 \oplus 1) \oplus (\mu_2 \oplus 1)) \oplus 1 \\ &= \mu_1 \oplus \mu_2 \oplus 1.\end{aligned}$$

Remark 1.7 The Cartesian product[3] of two functions $\Phi : \mathbf{B}^n \longrightarrow \mathbf{B}^p$, $\Psi : \mathbf{B}^m \longrightarrow \mathbf{B}^q$,

$$\mathbf{B}^n \ni \mu \mapsto \Phi(\mu) \in \mathbf{B}^p,$$

$$\mathbf{B}^m \ni \lambda \mapsto \Psi(\lambda) \in \mathbf{B}^q$$

is either of $\Phi \times \Psi : \mathbf{B}^n \times \mathbf{B}^m \longrightarrow \mathbf{B}^p \times \mathbf{B}^q$,

$$\mathbf{B}^n \times \mathbf{B}^m \ni (\mu, \lambda) \mapsto (\Phi(\mu), \Psi(\lambda)) \in \mathbf{B}^p \times \mathbf{B}^q$$

and $\Phi \times \Psi : \mathbf{B}^{n+m} \longrightarrow \mathbf{B}^{p+q}$,

$$\mathbf{B}^{n+m} \ni (\mu_1, \dots, \mu_n, \lambda_1, \dots, \lambda_m) \mapsto (\Phi_1(\mu_1, \dots, \mu_n), \dots, \Phi_p(\mu_1, \dots, \mu_n),$$
$$\Psi_1(\lambda_1, \dots, \lambda_m), \dots, \Psi_q(\lambda_1, \dots, \lambda_m)) \in \mathbf{B}^{p+q}.$$

Previously we have identified $\mathbf{B}^n \times \mathbf{B}^m$ and $\mathbf{B}^{n+m}$. We shall NOT use the Cartesian product of functions in the exposure for the sake of its conciseness, but many concepts to follow could be discussed in terms of Cartesian products.

1.3 Iterates

Notation 1.4 We denote with $\mathbf{N}$ the natural numbers set, $\mathbf{N} = \{0, 1, 2, \dots\}$.

Definition 1.8 The k-**iterate** of $\Phi : \mathbf{B}^n \longrightarrow \mathbf{B}^n$, where $k \in \mathbf{N}$, is the function $\Phi^{(k)} : \mathbf{B}^n \longrightarrow \mathbf{B}^n$,

$$\forall \mu \in \mathbf{B}^n, \Phi^{(k)}(\mu) = \begin{cases} \mu, \text{ if } k = 0, \\ (\underbrace{\Phi \circ \dots \circ \Phi}_{k})(\mu), \text{ if } k \geq 1 \end{cases}$$

and we agree that $(\underbrace{\Phi \circ \dots \circ \Phi}_{1}) = \Phi$.

Definition 1.9 The λ-**iterate** of $\Phi : \mathbf{B}^n \longrightarrow \mathbf{B}^n$, where $\lambda \in \mathbf{B}^n$, is the function $\Phi^\lambda : \mathbf{B}^n \longrightarrow \mathbf{B}^n$ defined by

$$\forall \mu \in \mathbf{B}^n, \forall i \in \{1, \dots, n\}, \Phi_i^\lambda(\mu) = \begin{cases} \mu_i, \text{ if } \lambda_i = 0, \\ \Phi_i(\mu), \text{ if } \lambda_i = 1. \end{cases}$$

Remark 1.8 Note that for both Definitions 1.8 and 1.9 the dimensions n of the domain and of the codomain of Φ, $\mathbf{B}^n$, must be equal.

---

3 See also Definitions A.2 and A.3, page 171.

Remark 1.9 The definition of $\Phi^{(k)}$ is intuitively related with the composition of a function with itself, these are the "synchronous" iterations of the function Φ. All the coordinates $\Phi_1, \dots, \Phi_n$ are computed at same the time, synchronously.

The definition of Φ^{λ} is intuitively related with the fact that the coordinates of Φ are not computed at the same time (only these coordinates Φ_i are computed for which $\lambda_i = 1$), i.e. $\Phi_1, \dots, \Phi_n$ are computed independently on each other. This gives the "asynchronous" iterations of Φ, when $\Phi, \Phi \circ \Phi, \dots$ are replaced by $\Phi^{\lambda}, \Phi^{\lambda'} \circ \Phi^{\lambda}, \dots$

Remark 1.10 Note that:

- synchronicity should be considered as a special case of asynchronicity, when all the coordinates $\Phi_1, \dots, \Phi_n$ are computed simultaneously;
- synchronicity means elegance and predictability;
- asynchronicity is more realistic in modeling.

Theorem 1.2 We have for all $k, p \in \mathbf{N}$, $\lambda, \nu \in \mathbf{B}^n$ and any $\mu \in \mathbf{B}^n$ that

$$(\Phi^{(k)})^{(p)}(\mu) = \Phi^{(kp)}(\mu), \tag{1.12}$$

$$(\Phi^{\lambda})^{\nu}(\mu) = \Phi^{\lambda\nu}(\mu). \tag{1.13}$$

Proof: (1.12): Case $k = 0$, then

$$(\Phi^{(0)})^{(p)} = (1_{\mathbf{B}^n})^{(p)} = 1_{\mathbf{B}^n} = \Phi^{(0)} = \Phi^{(0p)};$$

Case $p = 0$, then

$$(\Phi^{(k)})^{(0)} = 1_{\mathbf{B}^n} = \Phi^{(0)} = \Phi^{(k0)};$$

Case $k \geq 1, p \geq 1$, then

$$(\Phi^{(k)})^{(p)} = \underbrace{\underbrace{\Phi \circ \dots \circ \Phi}_{k} \circ \dots \circ \underbrace{\Phi \circ \dots \circ \Phi}_{k}}_{p} = \underbrace{\Phi \circ \dots \circ \Phi}_{kp}.$$

(1.13): We take $\lambda, \nu, \mu \in \mathbf{B}^n$ and $i \in \{1, \dots, n\}$ arbitrary and we get:

$$(\Phi^{\lambda})^{\nu}_i(\mu) = \begin{cases} \mu_i, & \text{if } \nu_i = 0, \\ \Phi^{\lambda}_i(\mu), & \text{if } \nu_i = 1 \end{cases} = \begin{cases} \mu_i, \text{ if } \nu_i = 0, \\ \mu_i, \text{ if } \lambda_i = 0 \text{ and } \nu_i = 1, \\ \Phi_i(\mu), \text{ if } \lambda_i = 1 \text{ and } \nu_i = 1 \end{cases}$$

$$= \Phi^{\lambda\nu}_i(\mu)$$

□

Theorem 1.3 We can write that

$$(\Phi^*)^{(k)}(\mu) = (\Phi^{(k)})^*(\mu), \tag{1.14}$$

$$(\Phi^*)^{\lambda}(\mu) = (\Phi^{\lambda})^*(\mu), \tag{1.15}$$

$$((\Phi^*)^{\lambda} \circ \ldots \circ (\Phi^*)^{\lambda'})(\mu) = (\Phi^{\lambda} \circ \ldots \circ \Phi^{\lambda'})^*(\mu) \tag{1.16}$$

hold for any $k \in \mathbf{N}$ and $\mu, \lambda, \ldots, \lambda' \in \mathbf{B}^n$.

Proof: We take $\mu \in \mathbf{B}^n$ arbitrary, fixed. The first statement is proved by induction on k. For $k = 0$, we have

$$(\Phi^*)^{(0)}(\mu) = 1_{\mathbf{B}^n}(\mu) = \mu = \overline{\overline{\mu}} = \overline{1_{\mathbf{B}^n}(\overline{\mu})} = (\Phi^{(0)})^*(\mu)$$

thus we suppose that (1.14) is true. We infer

$$\begin{aligned}(\Phi^*)^{(k+1)}(\mu) &= ((\Phi^*)^{(k)} \circ \Phi^*)(\mu) = ((\Phi^{(k)})^* \circ \Phi^*)(\mu) = (\Phi^{(k)})^*(\overline{\Phi(\overline{\mu})}) \\ &= \overline{\Phi^{(k)}(\Phi(\overline{\mu}))} = \overline{\Phi^{(k+1)}(\overline{\mu})} = (\Phi^{(k+1)})^*(\mu).\end{aligned}$$

We prove now (1.15) and let $\lambda \in \mathbf{B}^n$, $i \in \{1, \ldots, n\}$ arbitrary. We infer:

$$\begin{aligned}(\Phi^*)_i^{\lambda}(\mu) &= \begin{cases} \mu_i, & \text{if } \lambda_i = 0, \\ \Phi_i^*(\mu), & \text{if } \lambda_i = 1 \end{cases} = \begin{cases} \overline{\overline{\mu_i}}, & \text{if } \lambda_i = 0, \\ \overline{\Phi_i(\overline{\mu})}, & \text{if } \lambda_i = 1 \end{cases} = \overline{\Phi_i^{\lambda}(\overline{\mu})} \\ &= (\Phi^{\lambda})_i^*(\mu).\end{aligned}$$

In order to prove (1.16), we can write

$$\begin{aligned}&((\Phi^*)^{\lambda} \circ (\Phi^*)^{\lambda'})(\mu) = (\Phi^*)^{\lambda}((\Phi^*)^{\lambda'}(\mu)) = (\Phi^{\lambda})^*((\Phi^{\lambda'})^*(\mu)) \\ &= \overline{\Phi^{\lambda}(\overline{\overline{\Phi^{\lambda'}(\overline{\mu})}})} = \overline{\Phi^{\lambda}(\Phi^{\lambda'}(\overline{\mu}))} = \overline{(\Phi^{\lambda} \circ \Phi^{\lambda'})(\overline{\mu})} = (\Phi^{\lambda} \circ \Phi^{\lambda'})^*(\mu)\end{aligned}$$

and the general result is obtained by induction on the number of functions Φ^{λ} which are composed □

Theorem 1.4 The function Φ is given. For any $\mu \in \mathbf{B}^n$, we have the existence of $k' \in \mathbf{N}$ and $k'' > k'$ such that

$$\forall k \geq k', \Phi^{(k)}(\mu) = \Phi^{(k''-k'+k)}(\mu). \tag{1.17}$$

Proof: Indeed, the sequence $\Phi^{(k)}(\mu), k \in \mathbf{N}$ has finitely many values and $\exists k' \in \mathbf{N}, \exists k'' > k'$ such that $\Phi^{(k')}(\mu) = \Phi^{(k'')}(\mu)$. Property (1.17) is true □

Theorem 1.5 Let the function $\Phi : \mathbf{B}^n \longrightarrow \mathbf{B}^n$. We consider also that $\mu, \lambda \in \mathbf{B}^n, p, i_1, \ldots, i_p, q, j_1, \ldots, j_q \in \{1, \ldots, n\}$ are given.

(a) If

$$\Phi(\mu) = \mu \oplus \varepsilon^{i_1} \oplus \ldots \oplus \varepsilon^{i_p}, \tag{1.18}$$

then

$$\Phi^{\lambda}(\mu) = \mu \oplus \lambda_{i_1} \varepsilon^{i_1} \oplus \ldots \oplus \lambda_{i_p} \varepsilon^{i_p}; \tag{1.19}$$

(b) if

$$\Phi(\mu \oplus \varepsilon^{j_1} \oplus \ldots \oplus \varepsilon^{j_q}) = \mu, \tag{1.20}$$

then

$$\Phi^{\lambda}(\mu \oplus \varepsilon^{j_1} \oplus \ldots \oplus \varepsilon^{j_q}) = \mu \oplus \overline{\lambda_{j_1}} \varepsilon^{j_1} \oplus \ldots \oplus \overline{\lambda_{j_q}} \varepsilon^{j_q}; \tag{1.21}$$

(c) if

$$\Phi(\mu \oplus \varepsilon^{j_1} \oplus \ldots \oplus \varepsilon^{j_q}) = \mu \oplus \varepsilon^{i_1} \oplus \ldots \oplus \varepsilon^{i_p}, \tag{1.22}$$

then

$$\Phi^{\lambda}(\mu \oplus \varepsilon^{j_1} \oplus \ldots \oplus \varepsilon^{j_q}) = \mu \oplus \overline{\lambda_{j_1}} \varepsilon^{j_1} \oplus \ldots \oplus \overline{\lambda_{j_q}} \varepsilon^{j_q} \oplus \lambda_{i_1} \varepsilon^{i_1} \oplus \ldots \oplus \lambda_{i_p} \varepsilon^{i_p}. \tag{1.23}$$

Proof:

(a) The hypothesis states the truth of (1.18). For any $k \in \{1, \ldots, n\}$ we have

$$\Phi_k^{\lambda}(\mu) = \begin{cases} \mu_k, & k \in \{1, \ldots, n\} \backslash \{i_1, \ldots, i_p\}, \\ \mu_k, & k \in \{i_1, \ldots, i_p\} \text{ and } \lambda_k = 0, \\ \mu_k \oplus 1, & k \in \{i_1, \ldots, i_p\} \text{ and } \lambda_k = 1 \end{cases}$$

i.e. (1.19) is true.

(b) We have the truth of Eq. (1.20). For any $k \in \{1, \ldots, n\}$ we infer that

$$\Phi_k^{\lambda}(\mu \oplus \varepsilon^{j_1} \oplus \ldots \oplus \varepsilon^{j_q}) = \begin{cases} \mu_k, & k \in \{1, \ldots, n\} \backslash \{j_1, \ldots, j_q\}, \\ \mu_k \oplus 1, & k \in \{j_1, \ldots, j_q\} \text{ and } \lambda_k = 0, \\ \mu_k, & k \in \{j_1, \ldots, j_q\} \text{ and } \lambda_k = 1, \end{cases}$$

wherefrom the truth of (1.21) follows.

(c) From the truth of (1.22) we infer, $\forall k \in \{1,\dots,n\}$, that

$$\Phi_k^{\lambda}(\mu \oplus \varepsilon^{j_1} \oplus \dots \oplus \varepsilon^{j_q})$$
$$= \begin{cases} \mu_k, & k \in \{1,\dots,n\}\backslash(\{i_1,\dots,i_p\} \cup \{j_1,\dots,j_q\}), \\ \mu_k, & k \in \{i_1,\dots,i_p\}\backslash\{j_1,\dots,j_q\} \text{ and } \lambda_k = 0, \\ \mu_k \oplus 1, & k \in \{i_1,\dots,i_p\}\backslash\{j_1,\dots,j_q\} \text{ and } \lambda_k = 1, \\ \mu_k \oplus 1, & k \in \{j_1,\dots,j_q\}\backslash\{i_1,\dots,i_p\} \text{ and } \lambda_k = 0, \\ \mu_k, & k \in \{j_1,\dots,j_q\}\backslash\{i_1,\dots,i_p\} \text{ and } \lambda_k = 1, \\ \mu_k \oplus 1, & k \in \{i_1,\dots,i_p\} \cap \{j_1,\dots,j_q\} \end{cases}$$

and this shows the truth of (1.23) □

Remark 1.11 Theorem 1.3 has many applications, we show how it can be used in rewriting Theorem 1.4 and we give also a suggestion of using it in rewriting Theorem 1.5.

Theorem 1.6 Let $\mu \in \mathbf{B}^n$ arbitrary. If $k' \in \mathbf{N}$ and $k'' > k'$ fulfill (1.17), then

$$\forall k \geq k', \Phi^{*(k)}(\overline{\mu}) = \Phi^{*(k''-k'+k)}(\overline{\mu}).$$

Proof: We get $\forall k \geq k'$,

$$\Phi^{*(k)}(\overline{\mu}) \overset{\text{Theorem 1.3}}{=} \Phi^{(k)*}(\overline{\mu}) = \overline{\Phi^{(k)}(\overline{\overline{\mu}})} = \overline{\Phi^{(k)}(\mu)} = \overline{\Phi^{(k''-k'+k)}(\mu)}$$
$$= \overline{\Phi^{(k''-k'+k)}(\overline{\overline{\mu}})} = \Phi^{(k''-k'+k)*}(\overline{\mu}) \overset{\text{Theorem 1.3}}{=} \Phi^{*(k''-k'+k)}(\overline{\mu})$$ □

Theorem 1.7 For $\mu, \lambda \in \mathbf{B}^n, p, i_1,\dots,i_p, q, j_1,\dots,j_q \in \{1,\dots,n\}$,
(i) (1.18) is equivalent with

$$\Phi^*(\overline{\mu}) = \overline{\mu} \oplus \varepsilon^{i_1} \oplus \dots \oplus \varepsilon^{i_p}; \tag{1.24}$$

(ii) (1.24) implies

$$\Phi^{*\lambda}(\overline{\mu}) = \overline{\mu} \oplus \lambda_{i_1} \varepsilon^{i_1} \oplus \dots \oplus \lambda_{i_p} \varepsilon^{i_p}.$$

Proof: (i) (1.18) ⟹ (1.24) The computation is straight:

$$\Phi^*(\overline{\mu}) = \overline{\Phi(\overline{\overline{\mu}})} = \overline{\Phi(\mu)} \overset{(1.18)}{=} \overline{\mu \oplus \varepsilon^{i_1} \oplus \dots \oplus \varepsilon^{i_p}}$$
$$= \mu \oplus \varepsilon^{i_1} \oplus \dots \oplus \varepsilon^{i_p} \oplus (1,\dots,1)$$
$$= (\mu \oplus (1,\dots,1)) \oplus \varepsilon^{i_1} \oplus \dots \oplus \varepsilon^{i_p}$$
$$= \overline{\mu} \oplus \varepsilon^{i_1} \oplus \dots \oplus \varepsilon^{i_p}$$

and (1.24)⟹(1.18) is proved similarly.

(ii) We suppose that (1.24) is true, thus (1.18) is true also. Then Theorem 1.5 (a) shows the truth of (1.19) and we can write:

$$\Phi^{*\lambda}(\overline{\mu}) \overset{\text{Theorem 1.3}}{=} \Phi^{\lambda *}(\overline{\mu}) = \overline{\Phi^{\lambda}(\overline{\overline{\mu}})} = \overline{\Phi^{\lambda}(\mu)} \overset{(1.19)}{=} \overline{\mu \oplus \lambda_{i_1} \varepsilon^{i_1} \oplus \cdots \oplus \lambda_{i_p} \varepsilon^{i_p}}$$
$$= \mu \oplus \lambda_{i_1} \varepsilon^{i_1} \oplus \ldots \oplus \lambda_{i_p} \varepsilon^{i_p} \oplus (1, \ldots, 1)$$
$$= \overline{\mu} \oplus \lambda_{i_1} \varepsilon^{i_1} \oplus \ldots \oplus \lambda_{i_p} \varepsilon^{i_p}$$

□

1.4 State Portraits. Stable and Unstable Coordinates

Remark 1.12 The Boolean functions may be defined:

- by a formula, this was made at (1.8), (1.9), (1.10); in Definition 1.7, page 5, as Φ was supposed to be known, the new function Φ^* was defined by formula (1.11),
- by a table, we have done so at Table 1 in Preface,
- by a state portrait, if the dimensions of the domain and of the codomain are equal, and this is the topic of the present section. The possibility occurs because a state portrait corresponds to exactly one Boolean function.

Remark 1.13 The terminology of "state portrait," or rather "phase portrait," is specific to dynamical systems theory and gives the evolution in time of a system. In a binary context, the syntagms "state transition diagram" or "state transition graph" are preferred by many authors.

Definition 1.10 A **directed graph** is an ordered pair $G = (V, E)$, where $E \subset V \times V$. The elements of V are called **vertices, nodes,** or **points,** and the elements of E are called **arrows, directed edges,** or **directed arcs.**

Definition 1.11 The **state portrait** of the function $\Phi : \mathbf{B}^n \longrightarrow \mathbf{B}^n$ is the directed graph $G_\Phi = (V_\Phi, E_\Phi)$ defined by

$$V_\Phi = \mathbf{B}^n,$$
$$E_\Phi = \{(\mu, \mu') | \mu, \mu' \in \mathbf{B}^n, \mu \neq \mu' \text{and } \exists \lambda \in \mathbf{B}^n, \mu' = \Phi^{\lambda}(\mu)\}.$$

Remark 1.14 We give without proof the following result. The state portrait G_Φ defines the function $\Gamma : V_\Phi \longrightarrow V_\Phi$ by $\forall \mu \in V_\Phi$,

$$\Gamma(\mu) = \begin{cases} \mu \oplus \underset{i \in A_0 \cup \cdots \cup A_k}{\Xi} \varepsilon^i, & \text{if } \{(\mu', \mu'') | (\mu', \mu'') \in E_\Phi, \mu' = \mu\} \\ \quad = \{(\mu, \mu \oplus \underset{i \in A_0}{\Xi} \varepsilon^i), \ldots, (\mu, \mu \oplus \underset{i \in A_k}{\Xi} \varepsilon^i)\}, & \\ \mu, & \text{if } \{(\mu', \mu'') | (\mu', \mu'') \in E_\Phi, \mu' = \mu\} = \emptyset. \end{cases}$$

We have $\Gamma = \Phi$ and for each $\Omega : V_\Phi \longrightarrow V_\Phi$ with $\Omega \neq \Phi$, we get $G_\Phi \neq G_\Omega$.

Remark 1.15 In a state portrait, if an arrow exists from μ to μ' and $\mu' = \Phi^{\lambda}(\mu)$:

• we must have $\lambda \neq (0,\dots,0)$, otherwise we get $\mu' = \Phi^{(0,\dots,0)}(\mu) = \mu$, contradiction,

• we must have also a coordinate $i \in \{1,\dots,n\}$ with $\Phi_i(\mu) \neq \mu_i$, otherwise μ is a fixed point of Φ and $\forall i \in \{1,\dots,n\}, \forall \lambda \in \mathbf{B}^n$,

$$\mu_i' = \Phi_i^{\lambda}(\mu) = \begin{cases} \mu_i, \text{ if } \lambda_i = 0, \\ \Phi_i(\mu), \text{ if } \lambda_i = 1 \end{cases} = \begin{cases} \mu_i, \text{ if } \lambda_i = 0, \\ \mu_i, \text{ if } \lambda_i = 1 \end{cases} = \mu_i$$

representing a contradiction too. We use to underline $\underline{\mu_i}$ these coordinates of μ that fulfill $\Phi_i(\mu) \neq \mu_i$, since this way the drawing is more intuitive.

Definition 1.12 Let $\mu \in \mathbf{B}^n$. The coordinates $\mu_i, i \in \{1,\dots,n\}$ such that $\Phi_i(\mu) \neq \mu_i$ are called **unstable**, or **excited**, or **enabled** and the set of the unstable coordinates of μ is denoted with Φ_{μ}. The coordinates μ_i with $\Phi_i(\mu) = \mu_i$ are called **stable.** [4]

Remark 1.16 We obviously have, see Notation 1.3, page 4:

$$\Phi_{\mu} = \{i | i \in \{1,\dots,n\}, \mu_i \neq \Phi_i(\mu)\} = \mu \boxplus \Phi(\mu).$$

Example 1.5 We have drawn in Figure 1.1 the state portrait of the identity $1_{\mathbf{B}^2} : \mathbf{B}^2 \longrightarrow \mathbf{B}^2$. All the points $\mu \in \mathbf{B}^2$ are fixed points of $1_{\mathbf{B}^2}$, so there is no arrow and all the coordinates μ_i are stable.

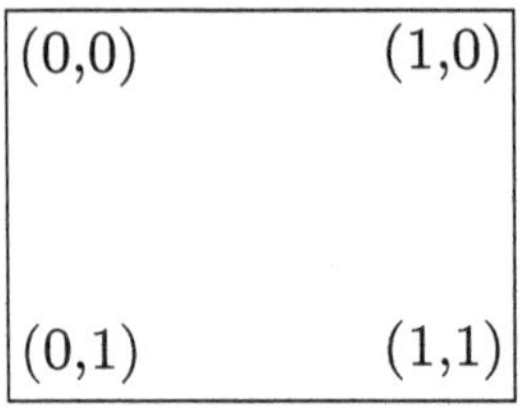

Figure 1.1 The state portrait of $1_{\mathbf{B}^2}$.

Vice versa, the only function $\Phi : \mathbf{B}^2 \longrightarrow \mathbf{B}^2$ whose state portrait is the one from Figure 1.1, i.e. that fulfills $\Phi(0,0) = (0,0), \Phi(1,0) = (1,0), \Phi(1,1) = (1,1)$, $\Phi(0,1) = (0,1)$, is the identity $1_{\mathbf{B}^2}$.

Example 1.6 The function $\Phi : \mathbf{B}^2 \longrightarrow \mathbf{B}^2$, $\forall \mu \in \mathbf{B}^2, \Phi(\mu_1, \mu_2) = (\mu_1 \cup \mu_2, \mu_1\mu_2)$ has its state portrait drawn in Figure 1.2. We see that Φ has three fixed points, $(0,0), (1,0), (1,1)$ with no unstable coordinate and no arrow starting from them. As $\Phi(0,1) = (1,0)$, the point $(0,1)$ has both coordinates unstable. There are also $2^2 - 1 = 3$ arrows starting from $(0,1)$, since $\Phi^{(1,0)}(0,1) = (1,1), \Phi^{(0,1)}(0,1) = (0,0)$ and $\Phi^{(1,1)}(0,1) = (1,0)$.

4 Both μ_i and i are called coordinates (of μ), but this creates no confusion.

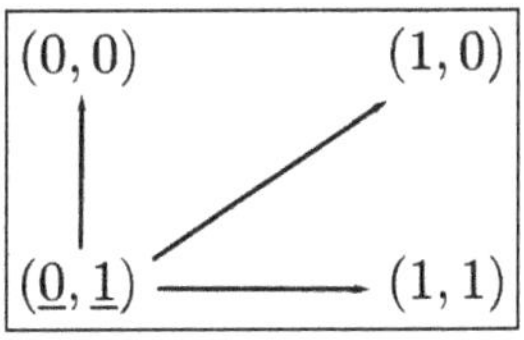

Figure 1.2 The state portrait of $\Phi(\mu_1, \mu_2) = (\mu_1 \cup \mu_2, \mu_1\mu_2)$.

Vice versa, the only function $\Phi : \mathbf{B}^2 \longrightarrow \mathbf{B}^2$ with the state portrait drawn in Figure 1.2, i.e. fulfilling $\Phi(0, 0) = (0, 0)$, $\Phi(1, 0) = (1, 0)$, $\Phi(1, 1) = (1, 1)$, $\Phi(0, 1) = (1, 0)$ is $\Phi(\mu_1, \mu_2) = (\mu_1 \cup \mu_2, \mu_1\mu_2)$.

Example 1.7 Figure 1.3 contains the state portrait of the constant function $\Phi : \mathbf{B}^2 \longrightarrow \mathbf{B}^2$, $\forall \mu \in \mathbf{B}^2, \Phi(\mu_1, \mu_2) = (1, 0)$. We notice that $2^2 - 1 = 3$ arrows start from $(0, 1)$, since it has two unstable coordinates, $2^1 - 1 = 1$ arrows start from $(0, 0), (1, 1)$, since they have one unstable coordinate and no arrows start (or $2^0 - 1 = 0$ arrows start) from $(1, 0)$, since it has no unstable coordinate, being a fixed point.

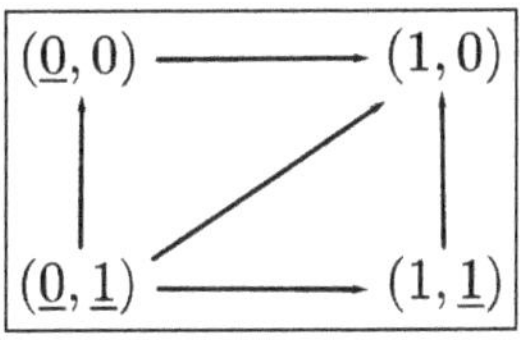

Figure 1.3 The state portrait of $\Phi(\mu_1, \mu_2) = (1, 0)$.

Vice versa, the only function $\Phi : \mathbf{B}^2 \longrightarrow \mathbf{B}^2$ with the state portrait from Figure 1.3 is $\Phi(\mu_1, \mu_2) = (1, 0)$.

Example 1.8 The function $\Phi : \mathbf{B}^2 \longrightarrow \mathbf{B}^2$, $\forall \mu \in \mathbf{B}^2, \Phi(\mu_1, \mu_2) = (\overline{\mu_1}, \overline{\mu_2})$ is characterized by the fact that all $\mu \in \mathbf{B}^2$ have both coordinates unstable and its state portrait was drawn in Figure 1.4.

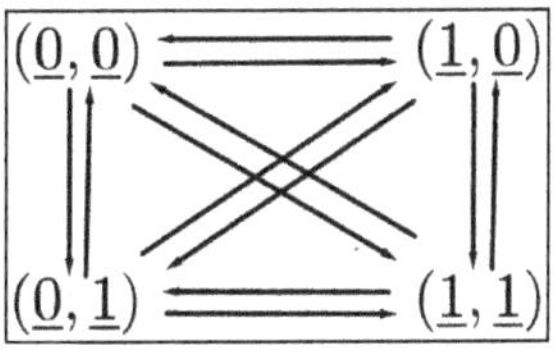

Figure 1.4 The state portrait of $\Phi(\mu_1, \mu_2) = (\overline{\mu_1}, \overline{\mu_2})$.

The only function with the state portrait drawn in Figure 1.4 is $\Phi(\mu_1, \mu_2) = (\overline{\mu_1}, \overline{\mu_2})$.

Example 1.9 We give in Figure 1.5 the state portrait of the dual function of $\Phi(\mu_1, \mu_2) = (\mu_1 \cup \mu_2, \mu_1\mu_2)$, i.e. $\Psi(\mu_1, \mu_2) = (\mu_1\mu_2, \mu_1 \cup \mu_2)$. We see that Figure 1.5 results from Figure 1.2 by complementing all the binary coordinates, while the arrows and the underlined coordinates remain the same.

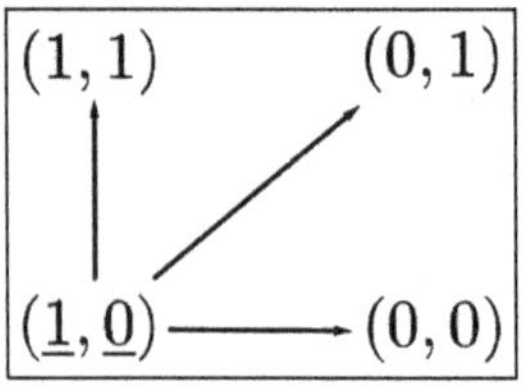

Figure 1.5 The state portrait of $\Psi(\mu_1, \mu_2) = (\mu_1\mu_2, \mu_1 \cup \mu_2)$.

1.5 Modeling the Asynchronous Circuits

Example 1.10 In Figure 1.6, we give the example of a circuit with two inputs ω_1, ω_2 and three logical gates, whose outputs were denoted with μ_1, μ_2, μ_3. The evolution is supposed to happen with ω_1, ω_2 constant; therefore, we get four functions $\Phi_\omega : \mathbf{B}^3 \longrightarrow \mathbf{B}^3, \omega \in \mathbf{B}^2$:

case $\omega_1 = 0, \omega_2 = 0 : \Phi_\omega(\mu_1, \mu_2, \mu_3) = (1, \mu_1\mu_3, \mu_2)$;
case $\omega_1 = 1, \omega_2 = 0 : \Phi_\omega(\mu_1, \mu_2, \mu_3) = (1, \mu_1\mu_3, 1)$;
case $\omega_1 = 0, \omega_2 = 1 : \Phi_\omega(\mu_1, \mu_2, \mu_3) = (0, \mu_1\mu_3, \mu_2)$;
case $\omega_1 = 1, \omega_2 = 1 : \Phi_\omega(\mu_1, \mu_2, \mu_3) = (0, \mu_1\mu_3, 1)$.

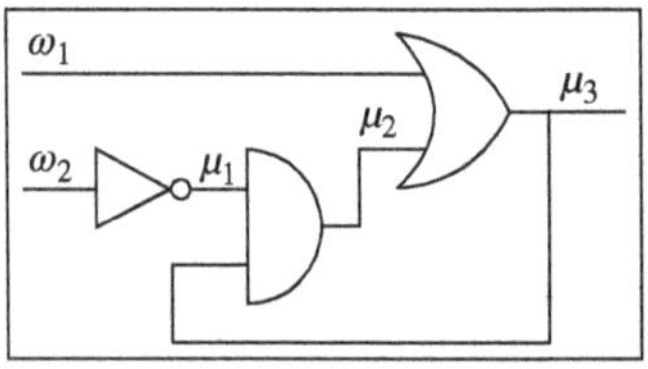

Figure 1.6 Circuit with three logical gates.

In the special case when $(\omega_1, \omega_2) = (1, 1)$, one can see in Figure 1.7 how the system stabilizes from any initial value of (μ_1, μ_2, μ_3) to $(0, 0, 1)$, which is a fixed point of Φ_ω.

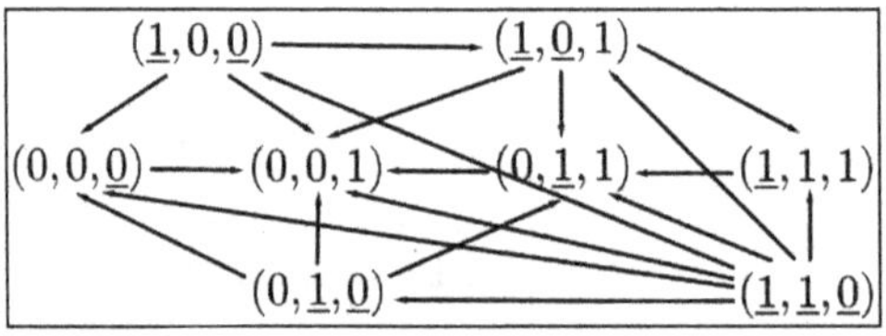

Figure 1.7 The state portrait of $\Phi_{(1,1)}(\mu_1, \mu_2, \mu_3) = (0, \mu_1\mu_3, 1)$.

1.6 Sequences of Sets

Definition 1.13 Let $A_k \subset \mathbf{B}^n, k \in \mathbf{N}$ a sequence of sets. If $k_1 \in \mathbf{N}$ exists such that $\forall k \geq k_1, A_k = A_{k_1}$, we say that the sequence is **convergent** and the set $A = A_{k_1}$ is called the **limit** of the sequence. We use the notation $A = \lim_{k\to\infty} A_k$.

Example 1.11 A descending sequence

$$A_0 \supset A_1 \supset A_2 \supset \ldots$$

is convergent and its limit is empty or not. This is analogue somehow with the descending sequences of real numbers that are bounded from below.

Example 1.12 Dually, an ascending sequence

$$A_0 \subset A_1 \subset A_2 \subset \ldots$$

of subsets of $\mathbf{B}^n$ is convergent. We have an analogy with the ascending sequences of real numbers that are bounded from above.

Remark 1.17 The limit in Example 1.11 coincides with the intersection $A_0 \cap A_1 \cap A_2 \cap \ldots$ and the limit in Example 1.12 coincides with the union $A_0 \cup A_1 \cup A_2 \cup \ldots$

1.7 Predecessors and Successors

Definition 1.14 Let $\Phi : \mathbf{B}^n \longrightarrow \mathbf{B}^n$ and we denote for any $\mu \in \mathbf{B}^n$:

$$\begin{aligned}
\mu^- &= \{\nu | \nu \in \mathbf{B}^n, \exists \lambda \in \mathbf{B}^n, \Phi^\lambda(\nu) = \mu\},\\
\mu^+ &= \{\Phi^\lambda(\mu) | \lambda \in \mathbf{B}^n\},\\
O^-(\mu) &= \{\nu | \nu \in \mathbf{B}^n, \exists \lambda \in \mathbf{B}^n, \ldots, \exists \lambda' \in \mathbf{B}^n, (\Phi^\lambda \circ \ldots \circ \Phi^{\lambda'})(\nu) = \mu\},\\
O^+(\mu) &= \{(\Phi^\lambda \circ \ldots \circ \Phi^{\lambda'})(\mu) | \lambda \in \mathbf{B}^n, \ldots, \lambda' \in \mathbf{B}^n\}.
\end{aligned}$$

The points $\nu \in \mu^-$ are called the **immediate predecessors** of μ, and the points $\nu \in \mu^+$ are called the **immediate successors** of μ; the points $\nu \in O^-(\mu)$ are the **predecessors** of μ, and the points $\nu \in O^+(\mu)$ are the **successors** of μ. If $\nu \in O^-(\mu)$, we say that μ is **accessible from** ν, and if $\nu \in O^+(\mu)$, we say that ν is **accessible from** μ.

Remark 1.18 If it will not be clear to which function Φ we refer, we shall use the notations $\mu^-_\Phi, \mu^+_\Phi, O^-_\Phi(\mu), O^+_\Phi(\mu)$.

Remark 1.19 Since $\Phi^{(0,\ldots,0)}(\mu) = \mu$, we infer that $\mu \in \mu^-$ and $\mu \in \mu^+$, in particular we have $\mu^- \neq \emptyset, \mu^+ \neq \emptyset$. On the other hand, $\mu^- \subset O^-(\mu), \mu^+ \subset O^+(\mu)$.

Example 1.13 (a) $\Phi(\mu) = \mu \Longrightarrow$

$$\mu^+ = \{\mu\},$$

(b) $\exists i \in \{1, \ldots, n\}, \Phi(\mu) = \mu \oplus \varepsilon^i \Longrightarrow$

$$\mu^+ = \{\mu, \mu \oplus \varepsilon^i\},$$

(c) $\exists i \in \{1,\dots,n\}, \exists j \in \{1,\dots,n\}, \Phi(\mu) = \mu \oplus \varepsilon^i \oplus \varepsilon^j \Longrightarrow$

$$\mu^+ = \{\mu, \mu \oplus \varepsilon^i, \mu \oplus \varepsilon^j, \mu \oplus \varepsilon^i \oplus \varepsilon^j\}$$

and the general case is obvious.

Theorem 1.8 We have $\forall \mu \in \mathbf{B}^n$,

$$\{\mu\} \cup \Phi^{-1}(\mu) \subset \mu^-, \tag{1.25}$$

$$\{\mu\} \cup \Phi^{-1}(\mu) \cup \Phi^{-1}(\Phi^{-1}(\mu)) \cup \dots \subset O^-(\mu), \tag{1.26}$$

$$\{\mu\} \cup \{\Phi(\mu)\} \subset \mu^+, \tag{1.27}$$

$$\{\mu\} \cup \{\Phi(\mu)\} \cup \{\Phi^{(2)}(\mu)\} \cup \dots \subset O^+(\mu). \tag{1.28}$$

Proof: (1.25). Let $\delta \in \{\mu\} \cup \Phi^{-1}(\mu)$ arbitrary. If $\delta = \mu$, then obviously $\delta \in \mu^-$; and if $\delta \in \Phi^{-1}(\mu)$, then $\lambda \in \mathbf{B}^n$ exists, $\lambda = (1,\dots,1)$, with $\Phi^\lambda(\delta) = \Phi(\delta) = \mu$ i.e. $\delta \in \mu^-$ □

Theorem 1.9 We have

$$O^-(\mu) = \mu^- \cup \bigcup_{\lambda \in \mu^-} \lambda^- \cup \bigcup_{\delta \in \bigcup_{\lambda \in \mu^-} \lambda^-} \delta^- \cup \dots, \tag{1.29}$$

$$O^+(\mu) = \mu^+ \cup \bigcup_{\lambda \in \mu^+} \lambda^+ \cup \bigcup_{\delta \in \bigcup_{\lambda \in \mu^+} \lambda^+} \delta^+ \cup \dots \tag{1.30}$$

Proof: (1.29). We denote $A_0 = \mu^-, A_1 = \bigcup_{\lambda \in A_0} \lambda^-, \dots, A_{k+1} = \bigcup_{\lambda \in A_k} \lambda^-, \dots$ and we show that

$$A_0 \subset A_1 \subset A_2 \subset \dots \tag{1.31}$$

Indeed, as $\mu \in \mu^-$, we can write that

$$A_1 = \bigcup_{\lambda \in \mu^-} \lambda^- \supset \bigcup_{\lambda \in \{\mu\}} \lambda^- = A_0, \tag{1.32}$$

$$A_2 = \bigcup_{\lambda \in A_1} \lambda^- \overset{(1.32)}{\supset} \bigcup_{\lambda \in A_0} \lambda^- = A_1$$

and (1.31) follows by induction. We are in the situation from Example 1.12, see also Remark 1.17, when some $k_1 \in \mathbf{N}$ exists with $A_{k_1} = A_{k_1+1} = A_{k_1+2} = \cdots$

We can write

$$\begin{aligned} \nu \in A_0 &\iff \exists \lambda \in \mathbf{B}^n, \Phi^{\lambda}(\nu) = \mu, \\ \nu \in A_1 &\iff \exists \lambda \in A_0, \nu \in \lambda^- \\ &\iff \exists \lambda \in \mathbf{B}^n, \exists \lambda' \in \mathbf{B}^n, \Phi^{\lambda'}(\lambda) = \mu \text{ and } \nu \in \lambda^- \\ &\iff \exists \lambda \in \mathbf{B}^n, \exists \lambda' \in \mathbf{B}^n, \Phi^{\lambda'}(\lambda) = \mu \text{ and } \exists \lambda'' \in \mathbf{B}^n, \Phi^{\lambda''}(\nu) = \lambda \\ &\iff \exists \lambda' \in \mathbf{B}^n, \exists \lambda'' \in \mathbf{B}^n, (\Phi^{\lambda'} \circ \Phi^{\lambda''})(\nu) = \mu \end{aligned}$$

and we can prove by induction on k that

$$\nu \in A_k \iff \exists \lambda^1 \in \mathbf{B}^n, \dots, \exists \lambda^{k+1} \in \mathbf{B}^n, (\Phi^{\lambda^1} \circ \dots \circ \Phi^{\lambda^{k+1}})(\nu) = \mu.$$

From the definition of $O^-(\mu)$, we get

$$O^-(\mu) = A_0 \cup A_1 \cup A_2 \cup \dots = A_{k_1}.$$

Equation (1.30). The proof is similar with the previous one, by replacing μ^- with μ^+, etc □

Theorem 1.10 The predecessors of the predecessors of μ are predecessors of μ, and the successors of the successors of μ are successors of μ :

$$\forall \mu' \in O^-(\mu), O^-(\mu') \subset O^-(\mu),$$

$$\forall \mu' \in O^+(\mu), O^+(\mu') \subset O^+(\mu).$$

Proof: In order to prove the first inclusion, let $\mu \in \mathbf{B}^n$, $\mu' \in O^-(\mu)$, and $\nu \in O^-(\mu')$ arbitrary, fixed. Then, $\lambda, \dots, \lambda', \omega, \dots, \omega' \in \mathbf{B}^n$ exist such that $(\Phi^{\lambda} \circ \dots \circ \Phi^{\lambda'})(\mu') = \mu, (\Phi^{\omega} \circ \dots \circ \Phi^{\omega'})(\nu) = \mu'$. We infer $(\Phi^{\lambda} \circ \dots \circ \Phi^{\lambda'} \circ \Phi^{\omega} \circ \dots \circ \Phi^{\omega'})(\nu) = \mu$, i.e. $\nu \in O^-(\mu)$ □

Remark 1.20 Let $\Phi, \Psi : \mathbf{B}^n \longrightarrow \mathbf{B}^n$ two functions. A nice question is if the statements $\Phi = \Psi$; $\forall \mu \in \mathbf{B}^n, \mu^-_{\Phi} = \mu^-_{\Psi}$; $\forall \mu \in \mathbf{B}^n, \mu^+_{\Phi} = \mu^+_{\Psi}$; $\forall \mu \in \mathbf{B}^n, O^-_{\Phi}(\mu) = O^-_{\Psi}(\mu)$; $\forall \mu \in \mathbf{B}^n, O^+_{\Phi}(\mu) = O^+_{\Psi}(\mu)$ are equivalent. A partial answer will be given in Remark 2.8, page 32.

Theorem 1.11 For any $\mu \in \mathbf{B}^n$, we can write

$$\begin{aligned} \overline{\mu}^-_{\Phi^*} &= \overline{\mu^-_{\Phi}}, \\ \overline{\mu}^+_{\Phi^*} &= \overline{\mu^+_{\Phi}}, \\ O^-_{\Phi^*}(\overline{\mu}) &= \overline{O^-_{\Phi}(\mu)}, \\ O^+_{\Phi^*}(\overline{\mu}) &= \overline{O^+_{\Phi}(\mu)}, \end{aligned}$$

where we have used the notations $\overline{X} = \{\overline{x} | x \in X\}, X \in \{\mu^-_{\Phi}, \mu^+_{\Phi}, O^-_{\Phi}(\mu), O^+_{\Phi}(\mu)\}$.

Proof: We prove the first equality:

$$\overline{\mu^-_{\Phi^*}} = \{\nu|\nu \in \mathbf{B}^n, \exists\lambda \in \mathbf{B}^n, (\Phi^*)^\lambda(\nu) = \overline{\mu}\}$$
$$\overset{(1.15)}{=} \{\nu|\nu \in \mathbf{B}^n, \exists\lambda \in \mathbf{B}^n, (\Phi^\lambda)^*(\nu) = \overline{\mu}\}$$
$$= \{\nu|\nu \in \mathbf{B}^n, \exists\lambda \in \mathbf{B}^n, \overline{\Phi^\lambda(\overline{\nu})} = \overline{\mu}\}$$
$$= \{\nu|\nu \in \mathbf{B}^n, \exists\lambda \in \mathbf{B}^n, \Phi^\lambda(\overline{\nu}) = \mu\}$$
$$= \{\overline{\nu}|\nu \in \mathbf{B}^n, \exists\lambda \in \mathbf{B}^n, \Phi^\lambda(\nu) = \mu\} = \overline{\mu^-_{\Phi}}.$$

The last equality is proved like this:

$$O^+_{\Phi^*}(\overline{\mu}) = \{((\Phi^*)^\lambda \circ \ldots \circ (\Phi^*)^{\lambda'})(\overline{\mu})|\lambda \in \mathbf{B}^n, \ldots, \lambda' \in \mathbf{B}^n\}$$
$$\overset{(1.16)}{=} \{(\Phi^\lambda \circ \ldots \circ \Phi^{\lambda'})^*(\overline{\mu})|\lambda \in \mathbf{B}^n, \ldots, \lambda' \in \mathbf{B}^n\}$$
$$= \{\overline{(\Phi^\lambda \circ \ldots \circ \Phi^{\lambda'})(\overline{\overline{\mu}})}|\lambda \in \mathbf{B}^n, \ldots, \lambda' \in \mathbf{B}^n\}$$
$$= \overline{\{(\Phi^\lambda \circ \ldots \circ \Phi^{\lambda'})(\overline{\overline{\mu}})|\lambda \in \mathbf{B}^n, \ldots, \lambda' \in \mathbf{B}^n\}} = \overline{O^+_{\Phi}(\mu)}$$

□

1.8 Source, Isolated Fixed Point, Transient Point, Sink

Definition 1.15 The function $\Phi : \mathbf{B}^n \longrightarrow \mathbf{B}^n$ is given. A point $\mu \in \mathbf{B}^n$ is called:

(a) *source:* if $\mu^- = \{\mu\}, \mu^+ \neq \{\mu\}$;
(b) *isolated fixed point:* if $\mu^- = \{\mu\}, \mu^+ = \{\mu\}$;
(c) *transient point:* if $\mu^- \neq \{\mu\}, \mu^+ \neq \{\mu\}$;
(d) *sink:* if $\mu^- \neq \{\mu\}, \mu^+ = \{\mu\}$.[5]

Example 1.14 We consider the function from Figure 1.8, where $(0,0,0)$, $(1,0,1)$ are sources, $(1,0,0), (1,1,0)$ are isolated fixed points, $(0,0,1), (0,1,1)$ are transient points and $(0,1,0), (1,1,1)$ are sinks. For example, we have:

$$(0,0,0)^- = \{(0,0,0)\},\ (0,0,0)^+ = \{(0,0,0), (0,1,0), (0,0,1), (0,1,1)\};$$
$$(1,0,0)^- = \{(1,0,0)\},\ (1,0,0)^+ = \{(1,0,0)\};$$
$$(0,0,1)^- = \{(0,0,0), (0,0,1)\},\ (0,0,1)^+ = \{(0,0,1), (0,1,1)\};$$
$$(0,1,0)^- = \{(0,0,0), (0,1,0)\},\ (0,1,0)^+ = \{(0,1,0)\}.$$

Remark 1.21 If a function $\Phi : \mathbf{B}^n \longrightarrow \mathbf{B}^n$ is given, any point $\mu \in \mathbf{B}^n$ is in exactly one of the situations (a)–(d) from Definition 1.15. We see that:

---

5 In [16], the garden-of-eden states are defined as "states that can arise only as initial states of the system and can never be dynamically generated during the course of the subsequent evolution" and this is equivalent here with $\mu^- = \{\mu\}$.

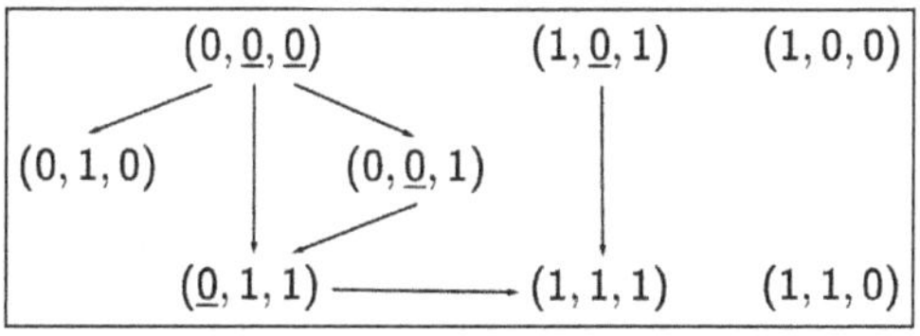

Figure 1.8 Example of sources, isolated fixed points, transient points, and sinks.

• the sources and the isolated fixed points, where $\mu^- = \{\mu\}$, fulfill the property that $\forall \lambda \in \mathbf{B}^n, (\Phi^\lambda)^{-1}(\mu) \in \{\emptyset, \{\mu\}\}$,

• the isolated fixed points and the sinks, where $\mu^+ = \{\mu\}$, satisfy either of $\forall \lambda \in \mathbf{B}^n, \Phi^\lambda(\mu) = \mu$ and $\Phi(\mu) = \mu$; in particular, the isolated fixed points are fixed points indeed.

Remark 1.22 Theorem 1.11 shows that $card(\overline{\mu}^-_{\Phi^*}) = card(\mu^-_\Phi)$ and $card(\overline{\mu}^+_{\Phi^*}) = card(\mu^+_\Phi)$, thus, μ is a source for Φ if and only if $\overline{\mu}$ is a source for Φ^*, μ is an isolated fixed point for Φ if and only if $\overline{\mu}$ is… Moreover, Eqs. (1.29) and (1.30) show that $\mu^- = \{\mu\} \Longleftrightarrow O^-(\mu) = \{\mu\}, \mu^+ = \{\mu\} \Longleftrightarrow O^+(\mu) = \{\mu\}$. Thus, Definition 1.15 may be expressed similarly by replacing μ^-, μ^+ with $O^-(\mu), O^+(\mu)$.

1.9 Translations

Notation 1.5 For $\tau \in \mathbf{B}^n$, we denote with $\theta^\tau : \mathbf{B}^n \longrightarrow \mathbf{B}^n$ the translation with $\tau : \forall \mu \in \mathbf{B}^n$,

$$\theta^\tau(\mu) = \mu \oplus \tau.$$

We also denote

$$\Theta_n = \{\theta^\tau | \tau \in \mathbf{B}^n\}.$$

Theorem 1.12 The set Θ_n is a commutative group relative to the composition of the functions, where:
(a) the neuter element is $\theta^{(0,\ldots,0)}$;
(b) $\forall \tau \in \mathbf{B}^n, \forall \tau' \in \mathbf{B}^n$ we have

$$\theta^\tau \circ \theta^{\tau'} = \theta^{\tau \oplus \tau'}; \tag{1.33}$$

(c) $\forall \tau \in \mathbf{B}^n$,

$$(\theta^\tau)^{-1} = \theta^\tau. \tag{1.34}$$

Proof: The statement (a) is obvious, since $\forall \mu \in \mathbf{B}^n$,

$$\theta^{(0,..,0)}(\mu) = \mu \oplus (0,\ldots,0) = \mu = 1_{\mathbf{B}^n}(\mu)$$

and $1_{\mathbf{B}^n}$ is the neuter element relative to the composition of the functions.
Let now $\tau, \tau', \mu \in \mathbf{B}^n$ arbitrary and fixed. We infer

$$(\theta^\tau \circ \theta^{\tau'})(\mu) = \theta^\tau(\theta^{\tau'}(\mu)) = \theta^\tau(\mu \oplus \tau') = (\mu \oplus \tau') \oplus \tau = \mu \oplus (\tau \oplus \tau')$$
$$= \theta^{\tau \oplus \tau'}(\mu),$$

thus (1.33) is true. If $\tau = \tau'$, then

$$\theta^\tau \circ \theta^\tau = \theta^{\tau \oplus \tau} = \theta^{(0,..,0)} = 1_{\mathbf{B}^n},$$

therefore (1.34) is also true. The commutativity of the group Θ_n is a consequence of (1.33):

$$\theta^\tau \circ \theta^{\tau'} = \theta^{\tau \oplus \tau'} = \theta^{\tau' \oplus \tau} = \theta^{\tau'} \circ \theta^\tau$$ □

2

Affine Spaces Defined by Two Points

The points $\mu, \lambda \in \mathbf{B}^n$ define the affine space $[\mu, \lambda] = \{\mu \oplus \underset{i\in A}{\Xi} \varepsilon^i | A \subset \mu \boxplus \lambda\}$ and the main reason of introducing it is given by the fact that, for $\Phi : \mathbf{B}^n \to \mathbf{B}^n$, we have $\mu^+ = [\mu, \Phi(\mu)]$. In other words, these spaces give the multiple possibilities of computation of a Boolean function. We have properties like $[\mu, \lambda] = [\lambda, \mu], [\mu, \mu] = \{\mu\}$, etc.

The functions $h : \mathbf{B}^n \to \mathbf{B}^n$ that fulfill $h([\mu, \lambda]) = [h(\mu), h(\lambda)]$ are said to be compatible with the affine structure of $\mathbf{B}^n$.

The Hamming distance between μ and λ is $d(\mu, \lambda) = card(\mu \boxplus \lambda)$. The function h is called Lipschitz if $\forall \mu \in \mathbf{B}^n, \forall \lambda \in \mathbf{B}^n$, $d(h(\mu), h(\lambda)) \leq d(\mu, \lambda)$, where $\leq$ is the order of the natural numbers. If h is compatible with the affine structure of $\mathbf{B}^n$, then it is Lipschitz. If h is bijective and compatible with the affine structure of $\mathbf{B}^n$, then $\forall \mu \in \mathbf{B}^n, \forall \lambda \in \mathbf{B}^n$, $d(h(\mu), h(\lambda)) = d(\mu, \lambda)$.

2.1 Definition

Notation 2.1 The points $\mu, \lambda \in \mathbf{B}^n$ are given. The following notations will be useful:

$$[\mu, \lambda] = \{\mu \oplus \underset{i\in A}{\Xi} \varepsilon^i | A \subset \mu \boxplus \lambda\},$$

$$[\mu, \lambda) = [\mu, \lambda] \setminus \{\lambda\},$$

$$(\mu, \lambda] = [\mu, \lambda] \setminus \{\mu\},$$

$$(\mu, \lambda) = [\mu, \lambda] \setminus \{\mu, \lambda\}.$$

Boolean Functions: Topics in Asynchronicity, First Edition. Serban E. Vlad.

Example 2.1 If $\mu = (0,0,1)$ and $\lambda = (0,1,0)$, we infer

$$\begin{aligned}[\mu,\lambda] &= \{(0,0,1) \oplus \underset{i\in A}{\Xi}\varepsilon^i | A \subset \{2,3\}\} \\ &= \{(0,0,1) \oplus \underset{i\in\emptyset}{\Xi}\varepsilon^i, (0,0,1) \oplus \underset{i\in\{2\}}{\Xi}\varepsilon^i, \\ &\quad (0,0,1) \oplus \underset{i\in\{3\}}{\Xi}\varepsilon^i, (0,0,1) \oplus \underset{i\in\{2,3\}}{\Xi}\varepsilon^i\} \\ &= \{(0,0,1), (0,1,1), (0,0,0), (0,1,0)\}.\end{aligned}$$

Remark 2.1 $[\mu,\lambda]$ may be interpreted as the set containing the points that are situated "between" μ and λ, some sort of "closed interval" or "closed line segment" with the ends μ and λ.

Remark 2.2 We get for $\mu, \lambda \in \mathbf{B}^n$:

$$[\mu,\lambda) = \{\mu \oplus \underset{i\in A}{\Xi}\varepsilon^i | A \subsetneq \mu \boxplus \lambda\},$$

$$(\mu,\lambda] = \{\mu \oplus \underset{i\in A}{\Xi}\varepsilon^i | \emptyset \subsetneq A \subset \mu \boxplus \lambda\},$$

$$(\mu,\lambda) = \{\mu \oplus \underset{i\in A}{\Xi}\varepsilon^i | \emptyset \subsetneq A \subsetneq \mu \boxplus \lambda\}.$$

Theorem 2.1 For any $\mu, \lambda \in \mathbf{B}^n$, we have that $[\mu,\lambda]$ is an affine space.

Proof: We prove first of all that the set

$$L = \{\underset{i\in A}{\Xi}\varepsilon^i | A \subset \mu \boxplus \lambda\}$$

is a linear space. As the product $(1.2)_{page\ 3}$ of the vectors with binary scalars is trivial, it is sufficient to show that L is closed relative to the modulo 2 sum of the vectors $(1.1)_{page\ 3}$. Let $\mu', \mu'' \in L$, $\mu' = \underset{i\in A'}{\Xi}\varepsilon^i$, $\mu'' = \underset{i\in A''}{\Xi}\varepsilon^i$ where $A', A'' \subset \mu \boxplus \lambda$[1]. We infer

$$\mu' \oplus \mu'' = \underset{i\in A'}{\Xi}\varepsilon^i \oplus \underset{i\in A''}{\Xi}\varepsilon^i = \underset{i\in A'\Delta A''}{\Xi}\varepsilon^i$$

and we have used the notation Δ for the symmetric difference of the sets. We have $A'\Delta A'' \subset \mu \boxplus \lambda$, therefore $\mu' \oplus \mu'' \in L$ follows.

We define now the function $\varphi : [\mu,\lambda] \times [\mu,\lambda] \to L, \forall (\mu',\mu'') \in [\mu,\lambda] \times [\mu,\lambda]$,

$$\varphi(\mu',\mu'') = \mu' \oplus \mu''.$$

1 The reasoning includes the possibility that $\mu = \lambda$, $\mu \boxplus \lambda = \emptyset$, $A' = A'' = \emptyset$, and $\mu' = \mu'' = (0,\ldots,0) \in \mathbf{B}^n$.

φ fulfills the next properties:

(i) $\forall \mu' \in [\mu, \lambda], \forall \mu'' \in [\mu, \lambda], \forall \mu''' \in [\mu, \lambda]$,

$$\varphi(\mu', \mu'') \oplus \varphi(\mu'', \mu''') = \varphi(\mu', \mu''').$$

Indeed, we have

$$\varphi(\mu', \mu'') \oplus \varphi(\mu'', \mu''') = \mu' \oplus \mu'' \oplus \mu'' \oplus \mu''' = \mu' \oplus \mu''' = \varphi(\mu', \mu''').$$

(ii) $\forall \mu' \in [\mu, \lambda], \forall \delta \in L, \exists! \mu'' \in [\mu, \lambda]$,

$$\varphi(\mu', \mu'') = \delta. \tag{2.1}$$

We have denoted with $\exists!$ the unique existence. For this, we put μ', μ'', δ under the form

$$\mu' = \mu \oplus \underset{i \in A'}{\Xi} \varepsilon^i,$$

$$\mu'' = \mu \oplus \underset{i \in A''}{\Xi} \varepsilon^i,$$

$$\delta = \underset{i \in B}{\Xi} \varepsilon^i$$

where $A', A'', B \subset \mu \boxplus \lambda$. Equation (2.1) means

$$\varphi(\mu', \mu'') = \mu' \oplus \mu'' = \mu \oplus \underset{i \in A'}{\Xi} \varepsilon^i \oplus \mu \oplus \underset{i \in A''}{\Xi} \varepsilon^i$$

$$= \underset{i \in A'}{\Xi} \varepsilon^i \oplus \underset{i \in A''}{\Xi} \varepsilon^i = \underset{i \in A' \Delta A''}{\Xi} \varepsilon^i = \underset{i \in B}{\Xi} \varepsilon^i = \delta$$

i.e.

$$A' \Delta A'' = B. \tag{2.2}$$

In equation (2.2), A', B are known and A'' is the unknown. The unique solution is given by

$$A'' = A' \Delta B,$$

therefore the unique μ'' that fulfills (2.1) is

$$\mu'' = \mu \oplus \underset{i \in A' \Delta B}{\Xi} \varepsilon^i.$$

The theorem is proved □

Definition 2.1 $[\mu, \lambda]$ is called the **affine space defined by** $\mu \in \mathbf{B}^n$ and $\lambda \in \mathbf{B}^n$.

2.2 Properties

Theorem 2.2 For any $\mu, \lambda, \tau \in \mathbf{B}^n$, the affine space $[\mu, \lambda]$ fulfills the following properties:

(a) $\nu \in [\mu, \lambda] \Longleftrightarrow \mu \boxplus \nu \subset \mu \boxplus \lambda$,

(b) $[\tau \oplus \mu, \tau \oplus \lambda] = \tau \oplus [\mu, \lambda]$, where we have denoted $\tau \oplus [\mu, \lambda] = \{\tau \oplus \nu | \nu \in [\mu, \lambda]\}$,

(c) $[\mu, \lambda] = [\lambda, \mu]$,

(d) $[\mu, \mu] = \{\mu\}$,
(e) $[\mu, \lambda] = [\mu, \lambda'] \Rightarrow \lambda = \lambda'$,
(f) $\nu \in [\mu, \lambda] \Longleftrightarrow [\nu, \lambda] \subset [\mu, \lambda]$.

Proof: In the proof we shall use several times Theorem 1.1, page 4.

(a) $\Longrightarrow$ If $\nu \in [\mu, \lambda]$, then $\nu = \mu \oplus \underset{i\in A}{\Xi}\varepsilon^i$, where $A \subset \mu \boxplus \lambda$. We get $\mu \oplus \nu = \underset{i\in\mu\boxplus\nu}{\Xi}\varepsilon^i = \underset{i\in A}{\Xi}\varepsilon^i$ i.e. $A = \mu \boxplus \nu \subset \mu \boxplus \lambda$.

$\Longleftarrow$ If $\mu \boxplus \nu \subset \mu \boxplus \lambda$, then $\nu = \mu \oplus \underset{i\in\mu\boxplus\nu}{\Xi}\varepsilon^i \in \{\mu \oplus \underset{i\in A}{\Xi}\varepsilon^i | A \subset \mu \boxplus \lambda\} = [\mu, \lambda]$.

(b) We can write

$$\begin{aligned}[\tau \oplus \mu, \tau \oplus \lambda] &= \{\tau \oplus \mu \oplus \underset{i\in A}{\Xi}\varepsilon^i | A \subset \tau \oplus \mu \boxplus \tau \oplus \lambda\}\\ &= \{\tau \oplus \mu \oplus \underset{i\in A}{\Xi}\varepsilon^i | A \subset \mu \boxplus \lambda\} = \{\tau \oplus \nu | \nu \in [\mu, \lambda]\}\\ &= \tau \oplus [\mu, \lambda].\end{aligned}$$

(c) We have

$$[\lambda, \mu] = \{\lambda \oplus \underset{i\in A}{\Xi}\varepsilon^i | A \subset \mu \boxplus \lambda\}.$$

We prove now the inclusion $[\mu, \lambda] \subset [\lambda, \mu]$ and let $\nu \in [\mu, \lambda]$ arbitrary. We can write

$$\nu = \mu \oplus \underset{i\in A}{\Xi}\varepsilon^i,$$

where $A \subset \mu \boxplus \lambda$, thus

$$\nu = \lambda \oplus \underset{i\in\mu\boxplus\lambda}{\Xi}\varepsilon^i \oplus \underset{i\in A}{\Xi}\varepsilon^i = \lambda \oplus \underset{i\in(\mu\boxplus\lambda)\Delta A}{\Xi}\varepsilon^i.$$

As $(\mu \boxplus \lambda)\Delta A = (\mu \boxplus \lambda)\backslash A \subset \mu \boxplus \lambda = \lambda \boxplus \mu$, we have obtained that $\nu \in [\lambda, \mu]$.

The inclusion $[\lambda, \mu] \subset [\mu, \lambda]$ is proved similarly.

(d) We can write that

$$\begin{aligned}[\mu, \mu] &= \{\mu \oplus \underset{i\in A}{\Xi}\varepsilon^i | A \subset \mu \boxplus \mu\}\\ &= \{\mu \oplus \underset{i\in A}{\Xi}\varepsilon^i | A \subset \emptyset\} = \{\mu \oplus \underset{i\in\emptyset}{\Xi}\varepsilon^i\} = \{\mu\}.\end{aligned}$$

(e) We have

$$\begin{aligned}&\lambda \in [\mu, \lambda'] \Rightarrow \mu \boxplus \lambda \subset \mu \boxplus \lambda',\\ &\lambda' \in [\mu, \lambda] \Rightarrow \mu \boxplus \lambda' \subset \mu \boxplus \lambda\end{aligned}$$

thus $\mu \boxplus \lambda = \mu \boxplus \lambda'$ and

$$\lambda = \mu \oplus \underset{i\in\mu\boxplus\lambda}{\Xi}\varepsilon^i = \mu \oplus \underset{i\in\mu\boxplus\lambda'}{\Xi}\varepsilon^i = \lambda'.$$

(f) We can write:

$$\nu \in [\mu, \lambda] \overset{(c)}{=} [\lambda, \mu] \overset{(a)}{\Longleftrightarrow} \lambda \boxplus \nu \subset \lambda \boxplus \mu$$

$$\Longleftrightarrow \{\lambda \oplus \underset{i\in A}{\Xi} \varepsilon^i | A \subset \lambda \boxplus \nu\} \subset \{\lambda \oplus \underset{i\in A}{\Xi} \varepsilon^i | A \subset \lambda \boxplus \mu\}$$

$$\Longleftrightarrow [\lambda, \nu] \subset [\lambda, \mu] \overset{(c)}{\Longleftrightarrow} [\nu, \lambda] \subset [\mu, \lambda] \qquad \square$$

Remark 2.3 A special form of the previous statement (b) exists for $\tau = (1, \dots, 1) \in \mathbf{B}^n$, namely

$$[\overline{\mu}, \overline{\lambda}] = \overline{[\mu, \lambda]}$$

with the notation $\overline{[\mu, \lambda]} = \{\overline{\nu} | \nu \in [\mu, \lambda]\}$.

Theorem 2.3 For any points $\mu, \nu, \lambda \in \mathbf{B}^n$ with $\nu \in [\mu, \lambda]$, we have

$$(\mu \boxplus \nu) \cap (\nu \boxplus \lambda) = \emptyset, \tag{2.3}$$

$$(\mu \boxplus \nu) \cup (\nu \boxplus \lambda) = \mu \boxplus \lambda. \tag{2.4}$$

Proof: We can write, using Theorem 2.2:

$$[\mu, \lambda] = \{\mu \oplus \underset{i\in H}{\Xi} \varepsilon^i | H \subset \mu \boxplus \lambda\} = \{\lambda \oplus \underset{i\in H}{\Xi} \varepsilon^i | H \subset \mu \boxplus \lambda\},$$

$$\nu = \mu \oplus \underset{i\in\mu\boxplus\nu}{\Xi} \varepsilon^i = \lambda \oplus \underset{i\in\nu\boxplus\lambda}{\Xi} \varepsilon^i,$$

where $\mu \boxplus \nu, \nu \boxplus \lambda \subset \mu \boxplus \lambda$.

We prove (2.3) by supposing against all reason the existence of an $i \in (\mu \boxplus \nu) \cap (\nu \boxplus \lambda)$, and this implies $\mu_i \oplus \nu_i = \lambda_i \oplus \nu_i = 1$. We infer $\mu_i \oplus \lambda_i = 0$, hence $i \notin \mu \boxplus \lambda$, representing a contradiction with the fact that $\mu \boxplus \nu, \nu \boxplus \lambda \subset \mu \boxplus \lambda$.

We prove (2.4). The inclusion $(\mu \boxplus \nu) \cup (\nu \boxplus \lambda) \subset \mu \boxplus \lambda$ results from $\mu \boxplus \nu \subset \mu \boxplus \lambda$ and $\nu \boxplus \lambda \subset \mu \boxplus \lambda$, so let us prove the inclusion $\mu \boxplus \lambda \subset (\mu \boxplus \nu) \cup (\nu \boxplus \lambda)$. For this, we take an arbitrary $i \in \mu \boxplus \lambda$, i.e. $\mu_i \oplus \lambda_i = 1$. Two possibilities exist.

If $\mu_i = \nu_i$, then $\nu_i \oplus \lambda_i = 1$ and $i \in \nu \boxplus \lambda \subset (\mu \boxplus \nu) \cup (\nu \boxplus \lambda)$.

If $\mu_i \neq \nu_i$, then $i \in \mu \boxplus \nu \subset (\mu \boxplus \nu) \cup (\nu \boxplus \lambda)$ $\square$

2.3 Functions that Are Compatible with the Affine Structure of B^n

Definition 2.2 A function $h : \mathbf{B}^n \to \mathbf{B}^n$ with the property that $\forall \mu \in \mathbf{B}^n$, $\forall \lambda \in \mathbf{B}^n$,

$$h([\mu, \lambda]) = [h(\mu), h(\lambda)] \tag{2.5}$$

is said to be **compatible with the affine structure of $\mathbf{B}^n$**. The set of the functions that are compatible with the affine structure of $\mathbf{B}^n$ is denoted by $Af(\mathbf{B}^n)$.

Theorem 2.4 The tuples $a, \tau \in \mathbf{B}^n$ are given and we define $h : \mathbf{B}^n \to \mathbf{B}^n$ by $\forall \mu \in \mathbf{B}^n$,

$$h(\mu) = \underset{i\in\{1,\ldots,n\}}{\Xi} a_i\mu_i\varepsilon^i \oplus \tau. \tag{2.6}$$

Then $h \in Af(\mathbf{B}^n)$.

Proof: We notice first that h has the property: for any $\mu \in \mathbf{B}^n$, $\lambda \in \mathbf{B}^n$,

$$\begin{aligned} h(\mu \oplus \lambda) &= \underset{i\in\{1,\ldots,n\}}{\Xi} a_i(\mu_i \oplus \lambda_i)\varepsilon^i \oplus \tau \\ &= \underset{i\in\{1,\ldots,n\}}{\Xi} a_i\mu_i\varepsilon^i \oplus \tau \oplus \underset{i\in\{1,\ldots,n\}}{\Xi} a_i\lambda_i\varepsilon^i \oplus \tau \oplus \tau \\ &= h(\mu) \oplus h(\lambda) \oplus \tau. \end{aligned} \tag{2.7}$$

On the other hand, for a better handling of the following computations, we use the notation $supp\ a = \{i | i \in \{1, \ldots, n\}, a_i = 1\}$, thus for any $A \subset \{1, \ldots, n\}$ we get

$$h(\underset{i\in A}{\Xi}\varepsilon^i) = \underset{i\in\{1,\ldots,n\}\cap A}{\Xi} a_i\varepsilon^i \oplus \tau = \underset{i\in supp\ a\cap A}{\Xi} \varepsilon^i \oplus \tau.$$

Let us take now $\mu \in \mathbf{B}^n$, $\lambda \in \mathbf{B}^n$ arbitrary, fixed. We can write:

$$\begin{aligned} h(\mu) \boxplus h(\lambda) &= \{i | i \in \{1, \ldots, n\}, h_i(\mu) \neq h_i(\lambda)\} \\ &= \{i | i \in \{1, \ldots, n\}, a_i\mu_i \oplus \tau_i \neq a_i\lambda_i \oplus \tau_i\} \\ &= \{i | i \in \{1, \ldots, n\}, a_i\mu_i \neq a_i\lambda_i\} \\ &= \{i | i \in \{1, \ldots, n\}, a_i = 1 \text{ and } \mu_i \neq \lambda_i\} \\ &= supp\ a \cap (\mu \boxplus \lambda). \end{aligned} \tag{2.8}$$

We infer:

$$\begin{aligned} h([\mu, \lambda]) &= h(\{\mu \oplus \underset{i\in A}{\Xi}\varepsilon^i | A \subset \mu \boxplus \lambda\}) = \{h(\mu \oplus \underset{i\in A}{\Xi}\varepsilon^i) | A \subset \mu \boxplus \lambda\} \\ &\overset{(2.7)}{=} \{h(\mu) \oplus h(\underset{i\in A}{\Xi}\varepsilon^i) \oplus \tau | A \subset \mu \boxplus \lambda\} \\ &= \{h(\mu) \oplus \underset{i\in supp\ a\cap A}{\Xi} \varepsilon^i \oplus \tau \oplus \tau | A \subset \mu \boxplus \lambda\} \\ &= \{h(\mu) \oplus \underset{i\in H}{\Xi}\varepsilon^i | H \subset supp\ a \cap (\mu \boxplus \lambda)\} \\ &\overset{(2.8)}{=} \{h(\mu) \oplus \underset{i\in H}{\Xi}\varepsilon^i | H \subset h(\mu) \boxplus h(\lambda)\} = [h(\mu), h(\lambda)] \end{aligned}$$

□

Theorem 2.5 We denote with $\sigma : \{1, \dots, n\} \to \{1, \dots, n\}$ a bijective function. Then $h : \mathbf{B}^n \to \mathbf{B}^n$ defined by $\forall \mu \in \mathbf{B}^n$,

$$h(\mu) = \underset{i\in\{1,\dots,n\}}{\Xi} \mu_{\sigma(i)}\varepsilon^i$$

is compatible with the affine structure of $\mathbf{B}^n$.

Proof: We notice first of all the additivity of $h : \forall \mu \in \mathbf{B}^n, \forall \lambda \in \mathbf{B}^n, h(\mu \oplus \lambda) = h(\mu) \oplus h(\lambda)$. Let us fix μ, λ, arbitrary. We obtain:

$$\begin{aligned} h(\mu) \boxplus h(\lambda) &= \{i | i \in \{1, \dots, n\}, h_i(\mu) \neq h_i(\lambda)\} \\ &= \{i | i \in \{1, \dots, n\}, \mu_{\sigma(i)} \neq \lambda_{\sigma(i)}\} \\ &= \{\sigma^{-1}(i) | i \in \{1, \dots, n\}, \mu_i \neq \lambda_i\} \\ &= \sigma^{-1}(\{i | i \in \{1, \dots, n\}, \mu_i \neq \lambda_i\}) = \sigma^{-1}(\mu \boxplus \lambda). \end{aligned}$$

We conclude:

$$\begin{aligned} h([\mu, \lambda]) &= h(\{\mu \oplus \underset{i\in A}{\Xi} \varepsilon^i | A \subset \mu \boxplus \lambda\}) = \{h(\mu \oplus \underset{i\in A}{\Xi} \varepsilon^i) | A \subset \mu \boxplus \lambda\} \\ &= \{h(\mu) \oplus \underset{i\in A}{\Xi} h(\varepsilon^i) | A \subset \mu \boxplus \lambda\} = \{h(\mu) \oplus \underset{i\in A}{\Xi} \varepsilon^{\sigma(i)} | A \subset \mu \boxplus \lambda\} \\ &= \{h(\mu) \oplus \underset{i\in\sigma^{-1}(A)}{\Xi} \varepsilon^i | A \subset \mu \boxplus \lambda\} \\ &= \{h(\mu) \oplus \underset{i\in\sigma^{-1}(A)}{\Xi} \varepsilon^i | \sigma^{-1}(A) \subset \sigma^{-1}(\mu \boxplus \lambda)\} \\ &= \{h(\mu) \oplus \underset{i\in H}{\Xi} \varepsilon^i | H \subset h(\mu) \boxplus h(\lambda)\} = [h(\mu), h(\lambda)] \end{aligned}$$

□

Example 2.2 The identity $1_{\mathbf{B}^n} : \mathbf{B}^n \to \mathbf{B}^n$ is compatible with the affine structure of $\mathbf{B}^n$. It represents the special case of (2.6) when $a_1 = \dots = a_n = 1$ and $\tau_1 = \dots = \tau_n = 0$.

Example 2.3 The translation $\theta^\tau : \mathbf{B}^n \to \mathbf{B}^n, \tau \in \mathbf{B}^n$ fulfills $\theta^\tau \in Af(\mathbf{B}^n)$. It is the special case of (2.6) when $a_1 = \dots = a_n = 1$.

Example 2.4 The constant function $h : \mathbf{B}^n \to \mathbf{B}^n$ equal with $\tau \in \mathbf{B}^n$ is compatible with the affine structure of $\mathbf{B}^n$. This is the special case of (2.6) when $a_1 = \dots = a_n = 0$.

Example 2.5 The function $h : \mathbf{B}^3 \to \mathbf{B}^3$ defined by $h(\mu_1, \mu_2, \mu_3) = (0, \mu_2 \oplus \mu_3, \mu_2 \oplus \mu_3)$ does not belong to $Af(\mathbf{B}^3)$ (to be compared with (2.6)). For this, we take $\mu = (0, 0, 0)$, $\lambda = (0, 1, 1)$ and we see that:

$$\begin{aligned} [(0,0,0),(0,1,1)] &= \{(0,0,0),(0,1,0),(0,0,1),(0,1,1)\}, \\ h([(0,0,0),(0,1,1)]) &= \{(0,0,0),(0,1,1)\}, \\ [h(0,0,0),h(0,1,1)] &= [(0,0,0),(0,0,0)] = \{(0,0,0)\}. \end{aligned}$$

Remark 2.4 If $h, g \in Af(\mathbf{B}^n)$, then $h \circ g \in Af(\mathbf{B}^n)$, meaning that $Af(\mathbf{B}^n)$ is a unitary semigroup relative to the composition of the functions.

Theorem 2.6 If h is bijective and $h \in Af(\mathbf{B}^n)$, then $\forall \mu \in \mathbf{B}^n, \forall \lambda \in \mathbf{B}^n$,

$$h((\mu, \lambda)) = (h(\mu), h(\lambda)).$$

Proof: Let $\mu \in \mathbf{B}^n$ and $\lambda \in \mathbf{B}^n$ arbitrary, for which (2.5) is true, thus $h((\mu, \lambda)) \subset [h(\mu), h(\lambda)]$. The statement of the theorem is false if $\omega \in (\mu, \lambda)$ exists such that $h(\omega) = h(\mu)$ or $h(\omega) = h(\lambda)$ holds. In the first case, for example, the fact that $\omega \neq \mu$ and the bijectivity of h give a contradiction. □

Theorem 2.7 If h is bijective and $h \in Af(\mathbf{B}^n)$, then $h^{-1} \in Af(\mathbf{B}^n)$.

Proof: We fix $\mu \in \mathbf{B}^n$, $\lambda \in \mathbf{B}^n$ arbitrary. We can write

$$[\mu, \lambda] = [h(h^{-1}(\mu)), h(h^{-1}(\lambda))] = h([h^{-1}(\mu), h^{-1}(\lambda)]),$$

thus

$$h^{-1}([\mu, \lambda]) = [h^{-1}(\mu), h^{-1}(\lambda)]$$ □

2.4 The Hamming Distance. Lipschitz Functions

Definition 2.3 We define $d : \mathbf{B}^n \times \mathbf{B}^n \to \{0, 1, \dots, n\}$ by $\forall \mu \in \mathbf{B}^n, \forall \lambda \in \mathbf{B}^n$,

$$d(\mu, \lambda) = card(\mu \boxplus \lambda).$$

d is called the **Hamming distance** between μ and λ.

Remark 2.5 For any $\tau \in \mathbf{B}^n$, we can write that

$$d(\tau \oplus \mu, \tau \oplus \lambda) = card(\tau \oplus \mu \boxplus \tau \oplus \lambda) = card(\mu \boxplus \lambda) = d(\mu, \lambda).$$

In particular, $\tau = (1, \dots, 1)$ gives $d(\overline{\mu}, \overline{\lambda}) = d(\mu, \lambda)$.

Theorem 2.8 For any $\mu \in \mathbf{B}^n, \lambda \in \mathbf{B}^n$, we have

$$card([\mu, \lambda]) = 2^{d(\mu, \lambda)}.$$

Proof: We infer for arbitrary μ, λ:

$$\begin{aligned} card([\mu, \lambda]) &= card(\{\mu \oplus \underset{i \in A}{\Xi} \varepsilon^i | A \subset \mu \boxplus \lambda\}) \\ &= card(\{A | A \subset \mu \boxplus \lambda\}) = 2^{card(\mu \boxplus \lambda)} = 2^{d(\mu, \lambda)} \end{aligned}$$ □

Definition 2.4 The function $h : \mathbf{B}^n \to \mathbf{B}^n$ is called **Lipschitz** if $\forall \mu \in \mathbf{B}^n$, $\forall \lambda \in \mathbf{B}^n$,

$$d(h(\mu), h(\lambda)) \leq d(\mu, \lambda) \tag{2.9}$$

is true, where $\leq$ is the inequality of the natural numbers.

Example 2.6 Let $a, \tau \in \mathbf{B}^n$ and the function $h : \mathbf{B}^n \to \mathbf{B}^n$ defined by $\forall \mu \in \mathbf{B}^n, h(\mu) = \underset{i \in \{1,\ldots,n\}}{\Xi} a_i \mu_i \varepsilon^i \oplus \tau$. We have

$$\begin{aligned} d(h(\mu), h(\lambda)) &= card(h(\mu) \boxplus h(\lambda)) \overset{(2.8)}{=} card(supp\ a \cap (\mu \boxplus \lambda)) \\ &\leq card(\mu \boxplus \lambda) = d(\mu, \lambda), \end{aligned}$$

i.e. h is Lipschitz. The special cases when h is the identity, the translation with τ and the constant function equal with τ are Lipschitz too.

Theorem 2.9 We suppose that $h \in Af(\mathbf{B}^n)$.

(a) h is Lipschitz;

(b) if in addition h is bijective, then $\forall \mu \in \mathbf{B}^n, \forall \lambda \in \mathbf{B}^n$,

$$d(\mu, \lambda) = d(h(\mu), h(\lambda)). \tag{2.10}$$

Proof: Let $\mu \in \mathbf{B}^n$, $\lambda \in \mathbf{B}^n$ arbitrary.

(a) We have

$$card([\mu, \lambda]) \geq card(h([\mu, \lambda])). \tag{2.11}$$

We infer

$$\begin{aligned} 2^{d(\mu,\lambda)} &\overset{\text{Theorem 2.8}}{=} card([\mu, \lambda]) \overset{(2.11)}{\geq} card(h([\mu, \lambda])) \\ &\overset{(2.5)}{=} card([h(\mu), h(\lambda)]) \overset{\text{Theorem 2.8}}{=} 2^{d(h(\mu),h(\lambda))}, \end{aligned}$$

thus (2.9) is true.

(b) The bijectivity of h makes (2.11) be replaced by

$$card([\mu, \lambda]) = card(h([\mu, \lambda])),$$

thus

$$2^{d(\mu,\lambda)} = card([\mu, \lambda]) = card(h([\mu, \lambda])) = card([h(\mu), h(\lambda)]) = 2^{d(h(\mu),h(\lambda))}.$$

We get the truth of (2.10) □

Example 2.7 If $h = \theta^\tau$ (the translation with $\tau \in \mathbf{B}^n$), then (2.10) is satisfied, see Remark 2.5, page 28.

Theorem 2.10 Let $h : \mathbf{B}^n \to \mathbf{B}^n$ bijective and Lipschitz. For any $\mu \in \mathbf{B}^n$ and any $k, i_1, \dots, i_k \in \{1, \dots, n\}$, we have the existence of $i_1^h, \dots, i_k^h \in \{1, \dots, n\}$ with

$$h(\mu \oplus \varepsilon^{i_1} \oplus \dots \oplus \varepsilon^{i_k}) = h(\mu) \oplus \varepsilon^{i_1^h} \oplus \dots \oplus \varepsilon^{i_k^h}.$$

Proof: We fix μ arbitrary and we use the induction on k.

We take $k = 1$ first. From

$$d(h(\mu), h(\mu \oplus \varepsilon^{i_1})) \geq 1$$

since $h(\mu) \neq h(\mu \oplus \varepsilon^{i_1})$ and

$$d(h(\mu), h(\mu \oplus \varepsilon^{i_1})) \leq d(\mu, \mu \oplus \varepsilon^{i_1}) = 1,$$

we get that $d(h(\mu), h(\mu \oplus \varepsilon^{i_1})) = 1$, i.e. $i_1^h \in \{1, \dots, n\}$ exists with $h(\mu \oplus \varepsilon^{i_1}) = h(\mu) \oplus \varepsilon^{i_1^h}$.

We suppose that the statement is true for $k - 1$ and we prove it for k. From

$$d(h(\mu \oplus \varepsilon^{i_1} \oplus \dots \oplus \varepsilon^{i_{k-1}}), h(\mu \oplus \varepsilon^{i_1} \oplus \dots \oplus \varepsilon^{i_{k-1}} \oplus \varepsilon^{i_k})) \geq 1$$

since $h(\mu \oplus \varepsilon^{i_1} \oplus \dots \oplus \varepsilon^{i_{k-1}}) \neq h(\mu \oplus \varepsilon^{i_1} \oplus \dots \oplus \varepsilon^{i_{k-1}} \oplus \varepsilon^{i_k})$ and

$$\begin{aligned} &d(h(\mu \oplus \varepsilon^{i_1} \oplus \dots \oplus \varepsilon^{i_{k-1}}), h(\mu \oplus \varepsilon^{i_1} \oplus \dots \oplus \varepsilon^{i_{k-1}} \oplus \varepsilon^{i_k})) \\ &\leq d(\mu \oplus \varepsilon^{i_1} \oplus \dots \oplus \varepsilon^{i_{k-1}}, \mu \oplus \varepsilon^{i_1} \oplus \dots \oplus \varepsilon^{i_{k-1}} \oplus \varepsilon^{i_k}) = 1 \end{aligned}$$

we infer that $d(h(\mu \oplus \varepsilon^{i_1} \oplus \dots \oplus \varepsilon^{i_{k-1}}), h(\mu \oplus \varepsilon^{i_1} \oplus \dots \oplus \varepsilon^{i_{k-1}} \oplus \varepsilon^{i_k})) = 1$, i.e. $i_k^h \in \{1, \dots, n\}$ exists such that

$$\begin{aligned} h(\mu \oplus \varepsilon^{i_1} \oplus \dots \oplus \varepsilon^{i_{k-1}} \oplus \varepsilon^{i_k}) &= h(\mu \oplus \varepsilon^{i_1} \oplus \dots \oplus \varepsilon^{i_{k-1}}) \oplus \varepsilon^{i_k^h} \\ &= h(\mu) \oplus \varepsilon^{i_1^h} \oplus \dots \oplus \varepsilon^{i_{k-1}^h} \oplus \varepsilon^{i_k^h}. \end{aligned}$$

The indexes $i_1^h, \dots, i_k^h$ are distinct otherwise, without loosing the generality, we get

$$\begin{aligned} h(\mu \oplus \varepsilon^{i_1} \oplus \dots \oplus \varepsilon^{i_k}) &= h(\mu) \oplus \varepsilon^{i_1^h} \oplus \dots \oplus \varepsilon^{i_{k-2}^h} \\ &= h(\mu \oplus \varepsilon^{i_1} \oplus \dots \oplus \varepsilon^{i_{k-2}}), \end{aligned}$$

contradiction with the bijectivity of h □

Corollary 2.1 If $h : \mathbf{B}^n \to \mathbf{B}^n$ is bijective and Lipschitz, then $\forall \mu \in \mathbf{B}^n$, $\forall \lambda \in \mathbf{B}^n$,

$$d(h(\mu), h(\lambda)) = d(\mu, \lambda). \tag{2.12}$$

Proof: We take $\mu, \lambda \in \mathbf{B}^n$ arbitrary, fixed. If $\mu = \lambda$, then (2.12) is true under the form

$$d(h(\mu), h(\mu)) = 0 = d(\mu, \mu),$$

so that we can suppose from now that $\mu \neq \lambda$. Let $k, i_1, \ldots, i_k \in \{1, \ldots, n\}$ with the property that

$$\lambda = \mu \oplus \varepsilon^{i_1} \oplus \ldots \oplus \varepsilon^{i_k}.$$

We infer from Theorem 2.10 the existence of $i_1^h, \ldots, i_k^h \in \{1, \ldots, n\}$ with

$$h(\lambda) = h(\mu \oplus \varepsilon^{i_1} \oplus \ldots \oplus \varepsilon^{i_k}) = h(\mu) \oplus \varepsilon^{i_1^h} \oplus \ldots \oplus \varepsilon^{i_k^h}$$

and we can write

$$\begin{aligned} d(h(\mu), h(\lambda)) &= d(h(\mu), h(\mu) \oplus \varepsilon^{i_1^h} \oplus \ldots \oplus \varepsilon^{i_k^h}) \\ &= k = d(\mu, \mu \oplus \varepsilon^{i_1} \oplus \ldots \oplus \varepsilon^{i_k}) = d(\mu, \lambda) \end{aligned}$$ □

2.5 Affine Spaces of Successors

Theorem 2.11 Let $\Phi : \mathbf{B}^n \longrightarrow \mathbf{B}^n$. For any $\mu \in \mathbf{B}^n$,

(a) the following equivalence holds:

$$\mu^+ = \{\mu\} \Longleftrightarrow \Phi(\mu) = \mu \Longleftrightarrow O^+(\mu) = \{\mu\};$$

(b) we have

$$\mu^+ = [\mu, \Phi(\mu)]. \tag{2.13}$$

Proof: (a) We refer to the first equivalence.

$\Rightarrow$ If $\mu^+ = \{\mu\}$, then $\forall \lambda \in \mathbf{B}^n, \Phi^\lambda(\mu) = \mu$, thus $\Phi^{(1,\ldots,1)}(\mu) = \Phi(\mu) = \mu$.

$\Leftarrow$ We suppose that $\Phi(\mu) = \mu$. Then $\forall \lambda \in \mathbf{B}^n, \forall i \in \{1, \ldots, n\}$,

$$\Phi_i^\lambda(\mu) = \begin{cases} \mu_i, \text{ if } \lambda_i = 0, \\ \Phi_i(\mu), \text{ if } \lambda_i = 1 \end{cases} = \mu_i.$$

(b) For any $\lambda \in \mathbf{B}^n$ and any $i \in \{1, \ldots, n\}$,

$$\Phi_i^\lambda(\mu) = \begin{cases} \mu_i, \text{ if } \lambda_i = 0, \\ \Phi_i(\mu), \text{ if } \lambda_i = 1 \end{cases} = \mu_i \oplus \lambda_i(\mu_i \oplus \Phi_i(\mu)),$$

thus

$$\begin{aligned} \Phi^\lambda(\mu) &= \underset{i\in\{1,\ldots,n\}}{\Xi} \Phi_i^\lambda(\mu)\varepsilon^i = \underset{i\in\{1,\ldots,n\}}{\Xi} (\mu_i \oplus \lambda_i(\mu_i \oplus \Phi_i(\mu)))\varepsilon^i \\ &= \underset{i\in\{1,\ldots,n\}}{\Xi} \mu_i\varepsilon^i \oplus \underset{i\in\{1,\ldots,n\}}{\Xi} \lambda_i(\mu_i \oplus \Phi_i(\mu))\varepsilon^i \\ &= \mu \oplus \underset{i\in\{1,\ldots,n\}}{\Xi} \lambda_i(\mu_i \oplus \Phi_i(\mu))\varepsilon^i. \end{aligned}$$

We obtain:

$$\mu^+ = \{\Phi^\lambda(\mu)|\lambda \in \mathbf{B}^n\} = \{\mu \oplus \underset{i\in\{1,\ldots,n\}}{\Xi} \lambda_i(\mu_i \oplus \Phi_i(\mu))\varepsilon^i|\lambda \in \mathbf{B}^n\}$$

$$= \{\mu \oplus \underset{i\in\{1,\ldots,n\}\cap\{j|j\in\{1,\ldots,n\},\mu_j\oplus\Phi_j(\mu)=1\}}{\Xi} \lambda_i\varepsilon^i|\lambda \in \mathbf{B}^n\}$$

$$= \{\mu \oplus \underset{i\in\Phi_\mu}{\Xi} \lambda_i\varepsilon^i|\lambda \in \mathbf{B}^n\} = \{\mu \oplus \underset{i\in\{j|j\in\{1,\ldots,n\},\lambda_j=1\}\cap\Phi_\mu}{\Xi} \varepsilon^i|\lambda \in \mathbf{B}^n\}$$

$$= \{\mu \oplus \underset{i\in A\cap\Phi_\mu}{\Xi} \varepsilon^i|A \subset \{1,\ldots,n\}\} = \{\mu \oplus \underset{i\in B}{\Xi}\varepsilon^i|B \subset \Phi_\mu\}$$

$$= [\mu, \Phi(\mu)].$$

We have used the fact that when λ runs in $\mathbf{B}^n$, the set $A = \{j|j \in \{1,\ldots,n\}, \lambda_j = 1\}$ runs in the subsets of $\{1,\ldots,n\}$ and the set $B = A \cap \Phi_\mu$ runs in the set of the subsets of Φ_μ □

Remark 2.6 The three statements of Theorem 2.11(a) are also equivalent with the statements $\overline{\mu}^+_{\Phi^*} = \{\overline{\mu}\}, \Phi^*(\overline{\mu}) = \overline{\mu}, O^+_{\Phi^*}(\overline{\mu}) = \{\overline{\mu}\}$.

Remark 2.7 The statement of Theorem 2.11(b), together with Theorem 1.11, page 17 give:

$$\overline{[\mu, \Phi(\mu)]} = \overline{\mu^+_\Phi} = \overline{\mu}^+_{\Phi^*} = [\overline{\mu}, \Phi^*(\overline{\mu})].$$

Remark 2.8 Given $\Psi : \mathbf{B}^n \longrightarrow \mathbf{B}^n$, the statements $\Phi = \Psi$ and $\forall \mu \in \mathbf{B}^n, \mu^+_\Phi = \mu^+_\Psi$ are equivalent, from (2.13) and Theorem 2.2 (e), page 23.

Remark 2.9 Let $\Phi : \mathbf{B}^n \longrightarrow \mathbf{B}^n$ now and the function $h : \mathbf{B}^n \longrightarrow \mathbf{B}^n$, which is compatible with the affine structure of $\mathbf{B}^n$. For any $\mu \in \mathbf{B}^n$, we have

$$h(\mu^+) \overset{\text{Theorem 2.11}}{=} h([\mu, \Phi(\mu)]) = [h(\mu), h(\Phi(\mu))].$$

Example 2.8 In Figure 2.1, we notice the existence of three fixed points $(1,0)$, $(0,1), (1,1) \in [(0,0), \Phi(0,0)] = \mathbf{B}^2$.

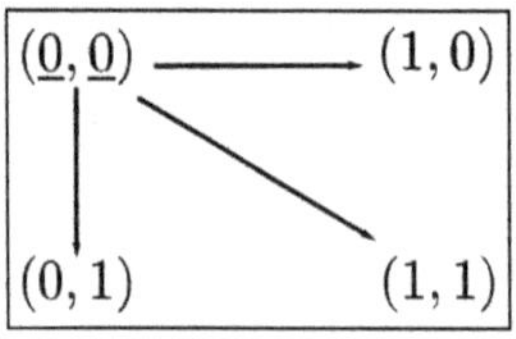

Figure 2.1 $[(0,0), \Phi(0,0)]$ contains three fixed points of $\Phi : (1,0), (0,1)$ and $(1,1)$.

Remark 2.10 By the use of (2.13), Eq. $(1.30)_{page\ 16}$ becomes:

$$O^+(\mu) = [\mu, \Phi(\mu)] \cup \bigcup_{\lambda \in [\mu, \Phi(\mu)]} [\lambda, \Phi(\lambda)] \cup \bigcup_{\delta \in \bigcup_{\lambda \in [\mu, \Phi(\mu)]} [\lambda, \Phi(\lambda)]} [\delta, \Phi(\delta)] \cup \dots$$

3

Morphisms

The category As whose objects are the functions $\Phi : \mathbf{B}^n \to \mathbf{B}^n$ is introduced in the first Appendix, where we show that it has finite products, and the morphisms that we work with belong, in fact, to that category.

The morphisms from $\Phi : \mathbf{B}^n \to \mathbf{B}^n$ to $\Psi : \mathbf{B}^n \to \mathbf{B}^n$ are couples of functions such that a diagram is commutative. The existence of a morphism from Φ to Ψ shows that Ψ behaves similarly with Φ. The composition of morphisms is a morphism.

The translations are special cases of morphisms. Defining the symmetry relative to translations allows, as an application, revisiting duality.

We present in this context how the fixed points, the predecessors and the successors are transferred from Φ to Ψ in the presence of a morphism.

3.1 Definition

Definition 3.1   We consider the functions $\Phi, \Psi : \mathbf{B}^n \to \mathbf{B}^n$. If $h, h' : \mathbf{B}^n \to \mathbf{B}^n$ exist such that $\forall v \in \mathbf{B}^n$, the diagram

$$
\begin{array}{ccc}
\mathbf{B}^n & \xrightarrow{\Phi^v} & \mathbf{B}^n \\
h \downarrow & & \downarrow h \\
\mathbf{B}^n & \xrightarrow{\Psi^{h'(v)}} & \mathbf{B}^n
\end{array}
$$

is commutative, then we denote $(h, h') : \Phi \to \Psi$ and we say that the **morphism** (h, h') is defined, from Φ to Ψ. If h, h' are both bijections, then (h, h') is called an **isomorphism** from Φ to Ψ. An isomorphism from Φ to Φ is called **automorphism**.

Notation 3.1   The sets of morphisms from Φ to Ψ, of isomorphisms from Φ to Ψ and of automorphisms of Φ are denoted by $Hom(\Phi, \Psi)$, $Iso(\Phi, \Psi)$ and $Aut(\Phi)$.

Boolean Functions: Topics in Asynchronicity, First Edition. Serban E. Vlad.
© 2019 John Wiley & Sons, Inc. Published 2019 by John Wiley & Sons, Inc.

Remark 3.1 The existence of a morphism from Φ to Ψ shows that Ψ behaves similarly with Φ, therefore we expect that there are properties of Φ which are transferred under some form to Ψ.

Remark 3.2 Here are some ways of generalizing the previous definition that we shall NOT use in the following:

• we can suppose the existence of three functions $h, h', h'' : \mathbf{B}^n \to \mathbf{B}^n$ (instead of two) such that $\forall \nu \in \mathbf{B}^n$, the diagram

$$\begin{array}{ccc} \mathbf{B}^n & \stackrel{\Phi^\nu}{\longrightarrow} & \mathbf{B}^n \\ h \downarrow & & \downarrow h'' \\ \mathbf{B}^n & \stackrel{\Psi^{h'(\nu)}}{\longrightarrow} & \mathbf{B}^n \end{array}$$

commutes;

• we can suppose that $\Phi : \mathbf{B}^n \to \mathbf{B}^n, \Psi : \mathbf{B}^m \to \mathbf{B}^m$, and $h, h' : \mathbf{B}^n \to \mathbf{B}^m$ (see Appendix A);

• we can have $\Phi : \mathbf{B}^m \times \mathbf{B}^n \to \mathbf{B}^n$ (the existence of an "input" $\omega \in \mathbf{B}^m$ in addition to the "state" $\mu \in \mathbf{B}^n$, if $\Phi = \Phi(\omega, \mu)$).

Theorem 3.1 For the functions $\Phi, \Psi : \mathbf{B}^n \to \mathbf{B}^n$, we have

$$(h, h') \in Hom(\Phi, \Psi) \Longleftrightarrow (h^*, h') \in Hom(\Phi^*, \Psi^*).$$

Proof: $\Longrightarrow$ Let $(h, h') \in Hom(\Phi, \Psi)$ and $\mu, \nu \in \mathbf{B}^n$ arbitrary. We can write, with the notation $\delta = \overline{\mu}$:

$$h^*(\Phi^{*\nu}(\mu)) = \overline{h(\overline{\Phi^{\nu*}(\overline{\delta})})} = \overline{h(\Phi^{\nu}(\delta))} = \overline{\Psi^{h'(\nu)}(h(\delta))} = \overline{\Psi^{h'(\nu)}(\overline{\overline{h(\overline{\overline{\overline{\delta}}})}})}$$

$$= \Psi^{h'(\nu)*}(\overline{h(\overline{\overline{\overline{\delta}}})}) = \Psi^{*h'(\nu)}(h^*(\overline{\delta})) = \Psi^{*h'(\nu)}(h^*(\mu)).$$

$\Longleftarrow$ We take $(h^*, h') \in Hom(\Phi^*, \Psi^*)$ and $\mu, \nu \in \mathbf{B}^n$ arbitrary. We have, with the notation $\delta = \overline{\mu}$:

$$h(\Phi^\nu(\mu)) = \overline{\overline{h(\overline{\overline{\Phi^{\nu}(\overline{\overline{\mu}})}})}} = \overline{h^*(\overline{\Phi^{\nu}(\overline{\delta})})} = \overline{h^*(\Phi^{\nu*}(\delta))} = \overline{h^*(\Phi^{*\nu}(\delta))}$$

$$= \overline{\Psi^{*h'(\nu)}(h^*(\delta))} = \overline{\Psi^{h'(\nu)*}(\overline{h(\overline{\delta})})} = \overline{\overline{\Psi^{h'(\nu)}(\overline{\overline{h(\mu)}})}} = \Psi^{h'(\nu)}(h(\mu)) \quad \square$$

3.2 Examples

Example 3.1 We have the automorphism $(1_{\mathbf{B}^n}, 1_{\mathbf{B}^n}) : \Phi \longrightarrow \Phi$.

Example 3.2 For $\Psi, h : \mathbf{B}^n \longrightarrow \mathbf{B}^n$ arbitrary and $h' : \mathbf{B}^n \longrightarrow \mathbf{B}^n$ defined by $\forall \mu \in \mathbf{B}^n, h'(\mu) = (0, \ldots, 0)$, we have that $(h, h') : 1_{\mathbf{B}^n} \longrightarrow \Psi$ is a morphism, since $\forall \mu \in \mathbf{B}^n, \forall \nu \in \mathbf{B}^n$,

$$\Psi^{h'(\nu)}(h(\mu)) = \Psi^{(0,\ldots,0)}(h(\mu)) = h(\mu) = h((1_{\mathbf{B}^n})^{\nu}(\mu)).$$

Example 3.3 Let $\mu' \in \mathbf{B}^n$ fixed, $\Phi, \Psi : \mathbf{B}^n \longrightarrow \mathbf{B}^n$ such that $\Psi(\mu') = \mu'$, $h : \mathbf{B}^n \longrightarrow \mathbf{B}^n$ defined by $\forall \mu \in \mathbf{B}^n, h(\mu) = \mu'$ and $h' : \mathbf{B}^n \longrightarrow \mathbf{B}^n$ arbitrary. We have the morphism $(h, h') : \Phi \longrightarrow \Psi$ since $\forall \mu \in \mathbf{B}^n, \forall \nu \in \mathbf{B}^n$,

$$\Psi^{h'(\nu)}(h(\mu)) = \Psi^{h'(\nu)}(\mu') = \mu' = h(\Phi^{\nu}(\mu)).$$

Example 3.4 We suppose that $\mu' \in \mathbf{B}^n$ is fixed, $\Phi, \Psi : \mathbf{B}^n \longrightarrow \mathbf{B}^n$ are arbitrary, $h : \mathbf{B}^n \longrightarrow \mathbf{B}^n$ is the constant function $\forall \mu \in \mathbf{B}^n, h(\mu) = \mu'$, and $h' : \mathbf{B}^n \longrightarrow \mathbf{B}^n$ is the null function $\forall \nu \in \mathbf{B}^n, h'(\nu) = (0, \ldots, 0)$. Then $(h, h') \in Hom(\Phi, \Psi) : \forall \mu \in \mathbf{B}^n, \forall \nu \in \mathbf{B}^n$,

$$\Psi^{h'(\nu)}(h(\mu)) = \Psi^{(0,\ldots,0)}(\mu') = \mu' = h(\Phi^{\nu}(\mu)).$$

Example 3.5 The functions $\Phi, \Psi : \mathbf{B}^2 \longrightarrow \mathbf{B}^2$ that are defined by $\forall \mu \in \mathbf{B}^2$,

$$\Phi(\mu) = (\mu_1 \cup \overline{\mu_2}, \mu_2),$$
$$\Psi(\mu) = (\mu_1 \cup \overline{\mu_2}, \overline{\mu_1} \cup \mu_2)$$

have their state portraits drawn in Figures 3.1 and 3.2. For $h, h' : \mathbf{B}^2 \longrightarrow \mathbf{B}^2$ given by $\forall \mu \in \mathbf{B}^2, h(\mu) = \mu$ and $h'(\mu) = (\mu_1, 0)$ we have that $(h, h') : \Phi \longrightarrow \Psi$ is a morphism, since $\forall \mu \in \mathbf{B}^2, \forall \nu \in \mathbf{B}^2$,

$$(h \circ \Phi^{\nu})(\mu) = (\overline{\nu_1}\mu_1 \cup \nu_1(\mu_1 \cup \overline{\mu_2}), \mu_2) = \Psi^{(\nu_1,0)}(\mu) = (\Psi^{h'(\nu)} \circ h)(\mu).$$

Remark 3.3 We infer from Example 3.1 that $Aut(\Phi) \neq \emptyset$ and $Hom(\Phi, \Psi)$, $Iso(\Phi, \Psi)$ may be empty.

Figure 3.1 The function $\Phi(\mu) = (\mu_1 \cup \overline{\mu_2}, \mu_2)$.

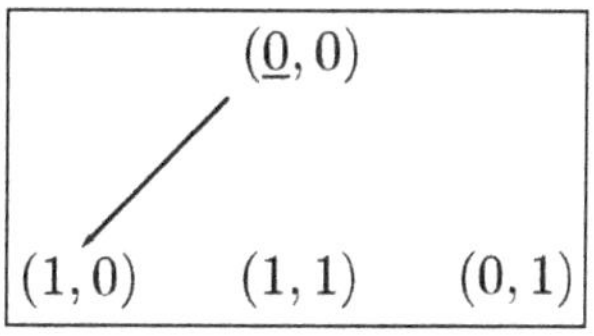

Figure 3.2 The function $\Psi(\mu) = (\mu_1 \cup \overline{\mu_2}, \overline{\mu_1} \cup \mu_2)$.

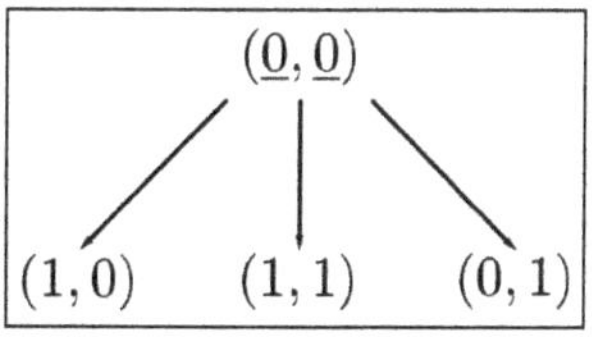

3.3 The Composition

Theorem 3.2 The functions $\Phi, \Psi, \Gamma, h, g, h', g' : \mathbf{B}^n \longrightarrow \mathbf{B}^n$ are given such that $(h, h') : \Phi \longrightarrow \Psi$, $(g, g') : \Psi \longrightarrow \Gamma$ are morphisms.

(a) $(g \circ h, g' \circ h') : \Phi \longrightarrow \Gamma$ is a morphism;

(b) if h, h' are bijections, then $(h^{-1}, h'^{-1}) : \Psi \longrightarrow \Phi$ is a morphism.

Proof: (a) This happens since $\forall \nu \in \mathbf{B}^n$, the diagrams

$$\begin{array}{ccc} \mathbf{B}^n & \overset{\Phi^\nu}{\longrightarrow} & \mathbf{B}^n \\ h \downarrow & & \downarrow h \\ \mathbf{B}^n & \overset{\Psi^{h'(\nu)}}{\longrightarrow} & \mathbf{B}^n \\ g \downarrow & & \downarrow g \\ \mathbf{B}^n & \overset{\Gamma^{(g' \circ h')(\nu)}}{\longrightarrow} & \mathbf{B}^n \end{array}$$

are all commutative: $\forall \mu \in \mathbf{B}^n$,

$$\begin{aligned} (\Gamma^{(g' \circ h')(\nu)} \circ (g \circ h))(\mu) &= ((\Gamma^{g'(h'(\nu))} \circ g) \circ h)(\mu) = ((g \circ \Psi^{h'(\nu)}) \circ h)(\mu) \\ &= (g \circ (\Psi^{h'(\nu)} \circ h))(\mu) \\ &= (g \circ (h \circ \Phi^\nu))(\mu) = ((g \circ h) \circ \Phi^\nu)(\mu). \end{aligned}$$

(b) For any $\nu \in \mathbf{B}^n$, we notice that

$$\Psi^{h'(\nu)} \circ h = h \circ \Phi^\nu$$

implies by multiplication with h^{-1} at the left and at the right

$$h^{-1} \circ \Psi^{h'(\nu)} = \Phi^\nu \circ h^{-1} = \Phi^{h'^{-1}(h'(\nu))} \circ h^{-1},$$

thus $(h^{-1}, h'^{-1}) \in Hom(\Psi, \Phi)$ □

Definition 3.2 The previous morphism $(g \circ h, g' \circ h')$ is by definition the **composition** of (g, g') and (h, h') and its usual notation is $(g, g') \circ (h, h')$.

Theorem 3.3 Let the functions $\Phi, \Psi : \mathbf{B}^n \longrightarrow \mathbf{B}^n$. The following statements are true:

(a) $\forall (g, g') \in Iso(\Psi, \Gamma), \forall (h, h') \in Iso(\Phi, \Psi), (g, g') \circ (h, h') \in Iso(\Phi, \Gamma)$;

(b) $Aut(\Phi)$ is a group relative to the composition of the morphisms, where the neuter element is $(1_{\mathbf{B}^n}, 1_{\mathbf{B}^n})$ and $\forall (h, h') \in Aut(\Phi)$, $(h, h')^{-1} = (h^{-1}, h'^{-1})$.

Proof: (a) If (g, g') and (h, h') are both isomorphisms, then $(g, g') \circ (h, h') = (g \circ h, g' \circ h')$ is an isomorphism: the composition of morphisms is a morphism from Theorem 3.2 and, in addition, the composition of bijections is a bijection.

(b) $(1_{\mathbf{B}^n}, 1_{\mathbf{B}^n})$ is the neuter element relative to the composition of the morphisms, since it belongs to $Aut(\Phi)$ and for any $(h, h') \in Aut(\Phi)$, we can write:

$$(h,h')\mathrm{o}(1_{\mathbf{B}^n},1_{\mathbf{B}^n}) = (h\mathrm{o}1_{\mathbf{B}^n},h'\mathrm{o}1_{\mathbf{B}^n}) = (h,h')$$
$$= (1_{\mathbf{B}^n}\mathrm{o}h, 1_{\mathbf{B}^n}\mathrm{o}h') = (1_{\mathbf{B}^n},1_{\mathbf{B}^n})\mathrm{o}(h,h').$$

In addition, (h^{-1}, h'^{-1}) exists, it is a morphism from Theorem 3.2 and

$$(h,h')\mathrm{o}(h^{-1},h'^{-1}) = (h\mathrm{o}h^{-1},h'\mathrm{o}h'^{-1}) = (1_{\mathbf{B}^n},1_{\mathbf{B}^n})$$
$$= (h^{-1}\mathrm{o}h, h'^{-1}\mathrm{o}h') = (h^{-1},h'^{-1})\mathrm{o}(h,h'),$$

therefore the inverse of (h, h') is (h^{-1}, h'^{-1}) □

3.4 A Fixed Point Property

Theorem 3.4 The functions $\Phi, \Psi : \mathbf{B}^n \longrightarrow \mathbf{B}^n$ are given, together with $\mu \in \mathbf{B}^n$ and $(h, h') \in Hom(\Phi, \Psi)$. If $\Phi(\mu) = \mu$, then

$$\forall \nu \in \mathbf{B}^n, \Psi^{h'(\nu)}(h(\mu)) = h(\mu),$$

and if h' is bijective, then $\Psi(h(\mu)) = h(\mu)$.

Proof: Indeed, for any $\nu \in \mathbf{B}^n$ we can write that

$$\Psi^{h'(\nu)}(h(\mu)) = h(\Phi^{\nu}(\mu)) = h(\mu),$$

and if h' is bijective, then $\nu' \in \mathbf{B}^n$ exists such that $h'(\nu') = (1, \dots, 1) \in \mathbf{B}^n$. We have:

$$\Psi(h(\mu)) = \Psi^{h'(\nu')}(h(\mu)) = h(\Phi^{\nu'}(\mu)) = h(\mu)$$ □

3.5 Symmetrical Functions Relative to Translations. Examples

Definition 3.3 The functions $\Phi, \Psi : \mathbf{B}^n \longrightarrow \mathbf{B}^n$ are called **symmetrical relative to the translation with** $\tau \in \mathbf{B}^n$, if $h' : \mathbf{B}^n \longrightarrow \mathbf{B}^n$ exists such that $(\theta^{\tau}, h') \in Iso(\Phi, \Psi)$ and $(\theta^{\tau}, h') \neq (1_{\mathbf{B}^n}, 1_{\mathbf{B}^n})$.

Definition 3.4 We say that the function $\Phi : \mathbf{B}^n \longrightarrow \mathbf{B}^n$ is **symmetrical relative to the translation with** τ, if $h' : \mathbf{B}^n \longrightarrow \mathbf{B}^n$ exists such that $(\theta^{\tau}, h') \in Aut(\Phi)$ and $(\theta^{\tau}, h') \neq (1_{\mathbf{B}^n}, 1_{\mathbf{B}^n})$.

Remark 3.4 The translations θ^τ are bijections, therefore the statements $(\theta^\tau, h') \in Iso(\Phi, \Psi)$, $(\theta^\tau, h') \in Aut(\Phi)$ make sense.

Example 3.6 The function $\Phi(\mu_1, \mu_2) = (\overline{\mu_2}, \mu_1)$ has its state portrait drawn in Figure 3.3. If we translate with $\tau = (1, 0)$ the points of $\mathbf{B}^2$ from Figure 3.3 we get Figure 3.4, where the state portrait of $\Psi(\mu_1, \mu_2) = (\mu_2, \overline{\mu_1})$ is drawn. Then Φ and Ψ are symmetrical relative to the translation with $(1, 0)$, *i.e.*$(\theta^{(1,0)}, 1_{\mathbf{B}^2}) \in Iso(\Phi, \Psi)$ takes place. Indeed, for any $\mu, \nu \in \mathbf{B}^2$ we can write:

$$\begin{aligned}(\theta^{(1,0)} \circ \Phi^\nu)(\mu) &= \theta^{(1,0)}(\overline{\nu_1}\mu_1 \oplus \nu_1\Phi_1(\mu), \overline{\nu_2}\mu_2 \oplus \nu_2\Phi_2(\mu))\\ &= ((\nu_1 \oplus 1)\mu_1 \oplus \nu_1(\mu_2 \oplus 1) \oplus 1, (\nu_2 \oplus 1)\mu_2 \oplus \nu_2\mu_1)\\ &= (\mu_1 \oplus \mu_1\nu_1 \oplus \mu_2\nu_1 \oplus \nu_1 \oplus 1, \mu_2 \oplus \mu_2\nu_2 \oplus \mu_1\nu_2)\end{aligned}$$

and on the other hand

$$\begin{aligned}(\Psi^\nu \circ \theta^{(1,0)})(\mu) &= \Psi^\nu(\mu_1 \oplus 1, \mu_2)\\ &= (\overline{\nu_1}(\mu_1 \oplus 1) \oplus \nu_1\Psi_1(\mu_1 \oplus 1, \mu_2),\\ &\quad \overline{\nu_2}\mu_2 \oplus \nu_2\Psi_2(\mu_1 \oplus 1, \mu_2))\\ &= ((\nu_1 \oplus 1)(\mu_1 \oplus 1) \oplus \nu_1\mu_2, (\nu_2 \oplus 1)\mu_2 \oplus \nu_2\mu_1)\\ &= (\mu_1 \oplus 1 \oplus \mu_1\nu_1 \oplus \nu_1 \oplus \nu_1\mu_2, \mu_2 \oplus \mu_2\nu_2 \oplus \mu_1\nu_2)\end{aligned}$$

i.e. $\theta^{(1,0)} \circ \Phi^\nu = \Psi^\nu \circ \theta^{(1,0)}$.

Example 3.7 We continue the reasoning with the function $\Phi(\mu_1, \mu_2) = (\overline{\mu_1}, \mu_2)$ from Figure 3.5. We translate each point of that figure with $(0, 1)$ and

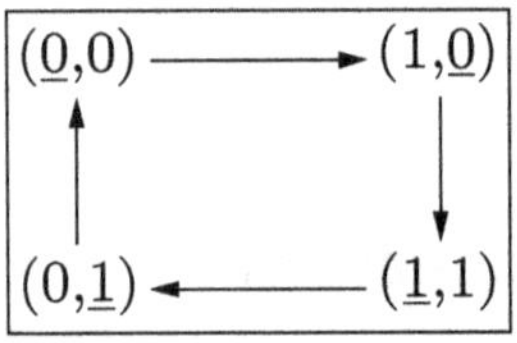

Figure 3.3 The function $\Phi(\mu_1, \mu_2) = (\overline{\mu_2}, \mu_1)$.

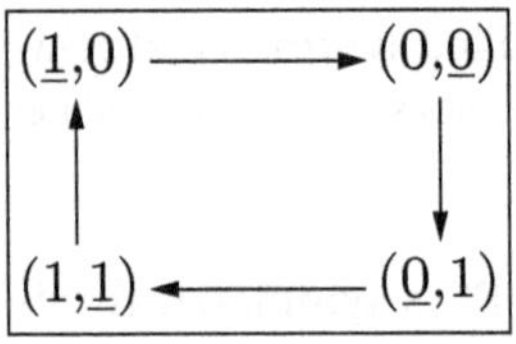

Figure 3.4 The function $\Psi(\mu_1, \mu_2) = (\mu_2, \overline{\mu_1})$.

we get the function from Figure 3.6, that coincides with Φ. We prove that $(\theta^{(0,1)}, 1_{\mathbf{B}^2}) \in Aut(\Phi)$ in the following way: $\forall \mu \in \mathbf{B}^2, \forall \nu \in \mathbf{B}^2$,

$$(\theta^{(0,1)} \circ \Phi^{\nu})(\mu) = \theta^{(0,1)}(\overline{\nu_1}\mu_1 \oplus \nu_1\overline{\mu_1}, \mu_2) = (\overline{\nu_1}\mu_1 \oplus \nu_1\overline{\mu_1}, \overline{\mu_2})$$

and

$$(\Phi^{\nu} \circ \theta^{(0,1)})(\mu) = \Phi^{\nu}(\mu_1, \overline{\mu_2}) = (\overline{\nu_1}\mu_1 \oplus \nu_1\overline{\mu_1}, \overline{\mu_2}).$$

We have obtained that $\theta^{(0,1)} \circ \Phi^{\nu} = \Phi^{\nu} \circ \theta^{(0,1)}$.

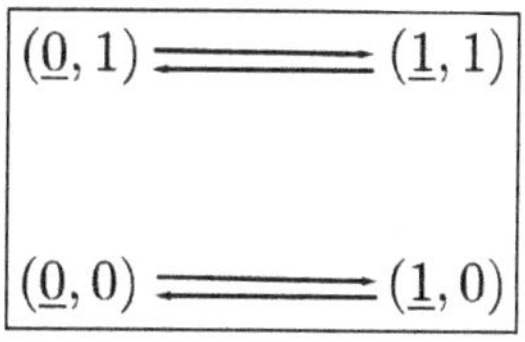

Figure 3.5 The function $\Phi(\mu_1, \mu_2) = (\overline{\mu_1}, \mu_2)$.

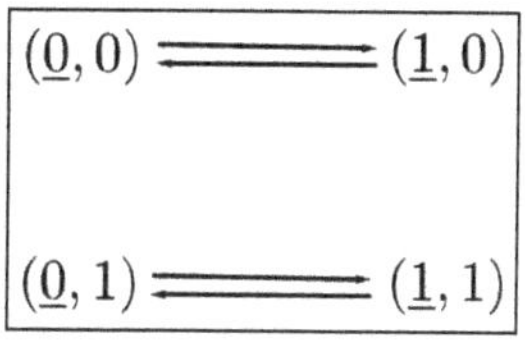

Figure 3.6 The function $\Phi(\mu_1, \mu_2) = (\overline{\mu_1}, \mu_2)$.

3.6 The Dual Functions Revisited

Remark 3.5 The translation with $(1, \dots, 1) \in \mathbf{B}^n$ has the meaning of complement, as we know. This allows revisiting the dual functions that prove to be isomorphic.

Theorem 3.5 Let the functions $\Phi, \Psi : \mathbf{B}^n \longrightarrow \mathbf{B}^n$; Ψ is the dual of Φ if and only if $(\theta^{(1,\dots,1)}, 1_{\mathbf{B}^n}) \in Iso(\Phi, \Psi)$.

Proof: We prove the if part first. From $(\theta^{(1,\dots,1)}, 1_{\mathbf{B}^n}) \in Iso(\Phi, \Psi)$ we can write

$$\theta^{(1,\dots,1)} \circ \Phi = \Psi \circ \theta^{(1,\dots,1)}. \tag{3.1}$$

We multiply (3.1) at the right with $\theta^{(1,\dots,1)}$ and we get

$$\Psi = \theta^{(1,\dots,1)} \circ \Phi \circ \theta^{(1,\dots,1)},$$

thus for any $\mu \in \mathbf{B}^n$ we infer

$$\Psi(\mu) = (\theta^{(1,\ldots,1)} \circ \Phi \circ \theta^{(1,\ldots,1)})(\mu) = \overline{\Phi(\overline{\mu})},$$

i.e. $\Psi = \Phi^*$.

We prove the only if part now. The hypothesis states that $\Psi = \Phi^*$ and let $i \in \{1,\ldots,n\}$, $\lambda, \mu \in \mathbf{B}^n$ arbitrary. We have

$$(\theta^{(1,\ldots,1)} \circ \Phi^{*\lambda})_i(\mu) = \overline{\Phi_i^{*\lambda}(\mu)} = \begin{cases} \overline{\mu_i}, & \text{if } \lambda_i = 0, \\ \overline{\Phi_i^*(\mu)}, & \text{if } \lambda_i = 1 \end{cases}$$

$$= \begin{cases} \overline{\mu_i}, & \text{if } \lambda_i = 0, \\ \overline{\overline{\Phi_i(\overline{\mu})}}, & \text{if } \lambda_i = 1 \end{cases}$$

$$= \begin{cases} \overline{\mu_i}, & \text{if } \lambda_i = 0, \\ \Phi_i(\overline{\mu}), & \text{if } \lambda_i = 1 \end{cases} = \Phi_i^\lambda(\overline{\mu}) = (\Phi^\lambda \circ \theta^{(1,\ldots,1)})_i(\mu),$$

in other words

$$\theta^{(1,\ldots,1)} \circ \Phi^{*\lambda} = \Phi^\lambda \circ \theta^{(1,\ldots,1)}. \tag{3.2}$$

We multiply (3.2) at the left and at the right with $\theta^{(1,\ldots,1)}$ and we obtain

$$\Phi^{*\lambda} \circ \theta^{(1,\ldots,1)} = \theta^{(1,\ldots,1)} \circ \Phi^\lambda,$$

therefore $(\theta^{(1,\ldots,1)}, 1_{\mathbf{B}^n}) \in Iso(\Phi, \Phi^*)$ □

3.7 Morphisms vs. Predecessors and Successors

Theorem 3.6 The functions $\Phi, \Psi : \mathbf{B}^n \longrightarrow \mathbf{B}^n$ are given, together with $\mu \in \mathbf{B}^n$ and $(h, h') \in Hom(\Phi, \Psi)$.

(a) We have

$$h(\mu_\Phi^-) \subset h(\mu)_\Psi^-, \tag{3.3}$$
$$h(\mu_\Phi^+) \subset h(\mu)_\Psi^+, \tag{3.4}$$
$$h(O_\Phi^-(\mu)) \subset O_\Psi^-(h(\mu)), \tag{3.5}$$
$$h(O_\Phi^+(\mu)) \subset O_\Psi^+(h(\mu)). \tag{3.6}$$

(b) If $(h, h') \in Iso(\Phi, \Psi)$, then

$$h(\mu_\Phi^-) = h(\mu)_\Psi^-, \tag{3.7}$$
$$h(\mu_\Phi^+) = h(\mu)_\Psi^+, \tag{3.8}$$
$$h(O_\Phi^-(\mu)) = O_\Psi^-(h(\mu)), \tag{3.9}$$
$$h(O_\Phi^+(\mu)) = O_\Psi^+(h(\mu)). \tag{3.10}$$

Proof: (a) We take $\lambda \in \mu_\Phi^-$ arbitrarily, i.e. $\nu \in \mathbf{B}^n$ exists such that $\Phi^\nu(\lambda) = \mu$. From

$$h(\mu) = h(\Phi^\nu(\lambda)) = \Psi^{h'(\nu)}(h(\lambda))$$

we infer that $h(\lambda) \in h(\mu)_\Psi^-$. Statement (3.3) is proved.

In order to prove (3.6), we take an arbitrary $\lambda \in O_\Phi^+(\mu)$, meaning the existence of $\nu \in \mathbf{B}^n, \ldots, \nu' \in \mathbf{B}^n$ with $\lambda = (\Phi^\nu \circ \ldots \circ \Phi^{\nu'})(\mu)$. Then

$$\begin{aligned} h(\lambda) &= h((\Phi^\nu \circ \ldots \circ \Phi^{\nu'})(\mu)) = (h \circ \Phi^\nu \circ \ldots \circ \Phi^{\nu'})(\mu) \\ &= (\Psi^{h'(\nu)} \circ h \circ \ldots \circ \Phi^{\nu'})(\mu) \ldots \\ &= (\Psi^{h'(\nu)} \circ \ldots \circ \Psi^{h'(\nu')} \circ h)(\mu) = (\Psi^{h'(\nu)} \circ \ldots \circ \Psi^{h'(\nu')})(h(\mu)), \end{aligned}$$

thus $h(\lambda) \in O_\Psi^+(h(\mu))$ and the truth of (3.6) follows.

(b) We prove

$$h(\mu)_\Psi^+ \subset h(\mu_\Phi^+) \tag{3.11}$$

and we take an arbitrary $\lambda' \in h(\mu)_\Psi^+$. This gives the existence of $\nu' \in \mathbf{B}^n$ with $\lambda' = \Psi^{\nu'}(h(\mu))$. The fact that h' is bijection shows the existence of $\nu \in \mathbf{B}^n$ with $\nu' = h'(\nu)$. We can write:

$$\lambda' = \Psi^{\nu'}(h(\mu)) = \Psi^{h'(\nu)}(h(\mu)) = h(\Phi^\nu(\mu))$$

where $\Phi^\nu(\mu) \in \mu_\Phi^+$, thus $\lambda' \in h(\mu_\Phi^+)$.Statement (3.11) is proved and, taking into account (3.4), we get the truth of (3.8).

We show that

$$O_\Psi^-(h(\mu)) \subset h(O_\Phi^-(\mu)). \tag{3.12}$$

Let $\lambda' \in O_\Psi^-(h(\mu))$ arbitrary, fixed. Then $\nu' \in \mathbf{B}^n, \ldots, \delta' \in \mathbf{B}^n$ exist such that $(\Psi^{\nu'} \circ \cdots \circ \Psi^{\delta'})(\lambda') = h(\mu)$. As h, h' are bijections, $\lambda, \nu, \ldots, \delta \in \mathbf{B}^n$ exist having the property that $\lambda' = h(\lambda), \nu' = h'(\nu), \ldots, \delta' = h'(\delta)$. We infer

$$\begin{aligned} h(\mu) &= (\Psi^{\nu'} \circ \ldots \circ \Psi^{\delta'})(\lambda') = (\Psi^{h'(\nu)} \circ \ldots \circ \Psi^{h'(\delta)})(h(\lambda)) \\ &= (\Psi^{h'(\nu)} \circ \ldots \circ \Psi^{h'(\delta)} \circ h)(\lambda) = (\Psi^{h'(\nu)} \circ \ldots \circ h \circ \Phi^\delta)(\lambda) = \ldots \\ &= (\Psi^{h'(\nu)} \circ h \circ \ldots \circ \Phi^\delta)(\lambda) = (h \circ \Phi^\nu \circ \ldots \circ \Phi^\delta)(\lambda) \\ &= h((\Phi^\nu \circ \ldots \circ \Phi^\delta)(\lambda)), \end{aligned}$$

i.e. $\mu = (\Phi^\nu \circ \ldots \circ \Phi^\delta)(\lambda)$. We get from here that $\lambda \in O_\Phi^-(\mu)$, thus $\lambda' = h(\lambda) \in h(O_\Phi^-(\mu))$. Statement (3.12) is true and, taking into account (3.5) also, we obtain the truth of (3.9) □

Corollary 3.1 We suppose that $(h, h') \in Iso(\Phi, \Psi)$ and let $\mu \in \mathbf{B}^n$.

(a) μ is a source of Φ if and only if $h(\mu)$ is a source of Ψ;

(b) μ is an isolated fixed point of Φ if and only if $h(\mu)$ is an isolated fixed point of Ψ;

(c) μ is a transient point of Φ if and only if $h(\mu)$ is a transient point of Ψ;
(d) μ is a sink of Φ if and only if $h(\mu)$ is a sink of Ψ.

Proof: h is bijective and we use Eqs. (3.7) and (3.8) that imply:

$$card(\mu_{\Phi}^{-}) = card(h(\mu_{\Phi}^{-})) = card(h(\mu)_{\Psi}^{-}),$$
$$card(\mu_{\Phi}^{+}) = card(h(\mu_{\Phi}^{+})) = card(h(\mu)_{\Psi}^{+})$$

□

Remark 3.6 To be compared the previous corollary with Remark 1.22, page 19.

4

Antimorphisms

The antimorphisms are defined like the morphisms, the antimorphisms from $\Phi : \mathbf{B}^n \to \mathbf{B}^n$ to $\Psi : \mathbf{B}^n \to \mathbf{B}^n$ are couples of functions such that a diagram is commutative. In contradistinction to the morphisms, the cause–effect sense of Φ and Ψ is inversed, suggesting this way time reversal. The existence of an antimorphism from Φ to Ψ shows that some properties of Φ are transferred to Ψ.

The composition of the antimorphisms is a morphism, and the composition of an antimorphism with a morphism is an antimorphism.

The antisymmetry relative to translations is discussed, as well as the action of the antimorphisms on fixed points, predecessors and successors.

4.1 Definition

Definition 4.1 Let us consider the functions $\Phi, \Psi : \mathbf{B}^n \to \mathbf{B}^n$ for which $h, h' : \mathbf{B}^n \to \mathbf{B}^n$ exist with the property that $\forall v \in \mathbf{B}^n$, the diagram

$$\begin{array}{ccc} \mathbf{B}^n & \stackrel{\Phi^v}{\longrightarrow} & \mathbf{B}^n \\ h \downarrow & & \downarrow h \\ \mathbf{B}^n & \stackrel{\Psi^{h'(v)}}{\longleftarrow} & \mathbf{B}^n \end{array}$$

is commutative. We denote then $(h, h')^\frown : \Phi \to \Psi$ and we say that the **antimorphism** $(h, h')^\frown$ is defined, from Φ to Ψ. If the functions h, h' are both bijections and $(h^{-1}, h'^{-1})^\frown : \Psi \longrightarrow \Phi$ is antimorphism, then $(h, h')^\frown$ is called **antiisomorphism** from Φ to Ψ and if $(h, h')^\frown : \Phi \to \Phi$ is antiisomorphism, then we call it **antiautomorphism**.

Notation 4.1 We denote the sets of the antimorphisms from Φ to Ψ, of the antiisomorphisms from Φ to Ψ and of the antiautomorphisms of Φ with $Hom^\frown(\Phi, \Psi)$, $Iso^\frown(\Phi, \Psi)$ and $Aut^\frown(\Phi)$.

Boolean Functions: Topics in Asynchronicity, First Edition. Serban E. Vlad.

Remark 4.1 The meaning of the existence of an antimorphism from Φ to Ψ is similar with that of the existence of a morphism, but with the cause-effect sense of Φ and Ψ inversed, suggesting that, when Φ and Ψ are computed (asynchronously), time flows in opposite senses. Reversing time is a topic in physics and its adaptation to asynchronicity (in this timeless framework) will be addressed later. If an antimorphism from Φ to Ψ exists, we expect that some properties of Φ be transferred under some form to Ψ.

Remark 4.2 Like in the case of the morphisms, see Remark 3.2, page 36, several possibilities exist of generalizing Definition 4.1.

Remark 4.3 For arbitrary Φ, Ψ any of $Hom^{\smile}(\Phi, \Psi), Iso^{\smile}(\Phi, \Psi), Aut^{\smile}(\Phi)$ can be empty.

Theorem 4.1 We have $(h, h') \in Hom^{\smile}(\Phi, \Psi) \Longleftrightarrow (h^*, h') \in Hom^{\smile}(\Phi^*, \Psi^*)$.

Proof: $\Longrightarrow$ Let $(h, h') \in Hom^{\smile}(\Phi, \Psi)$ and $\mu, \nu \in \mathbf{B}^n$ arbitrary. We denote $\delta = \overline{\mu}$. We have:

$$h^*(\mu) = \overline{h(\delta)} = \overline{\Psi^{h'(\nu)}(h(\Phi^\nu(\delta)))} = \overline{\Psi^{h'(\nu)}(\overline{\overline{h(\Phi^\nu(\delta))}})} = \Psi^{h'(\nu)*}(\overline{h(\overline{\overline{\Phi^\nu(\delta)}})})$$

$$= \Psi^{*h'(\nu)}(h^*(\overline{\Phi^\nu(\delta)})) = \Psi^{*h'(\nu)}(h^*(\Phi^{\nu*}(\bar{\delta}))) = \Psi^{*h'(\nu)}(h^*(\Phi^{*\nu}(\mu))).$$

$\Longleftarrow$ We take $(h^*, h') \in Hom^{\smile}(\Phi^*, \Psi^*)$, $\mu, \nu \in \mathbf{B}^n$ arbitrary and we denote $\delta = \overline{\mu}$. We can write:

$$h(\mu) = \overline{\overline{h(\overline{\overline{\mu}})}} = \overline{h^*(\delta)} = \overline{\Psi^{*h'(\nu)}(h^*(\Phi^{*\nu}(\delta)))} = \overline{\Psi^{h'(\nu)*}(h^*(\Phi^{\nu*}(\delta)))}$$

$$= \Psi^{h'(\nu)}(\overline{h^*(\Phi^{\nu*}(\delta))}) = \Psi^{h'(\nu)}(h(\overline{\Phi^{\nu*}(\delta)})) = \Psi^{h'(\nu)}(h(\Phi^\nu(\bar{\delta})))$$

$$= \Psi^{h'(\nu)}(h(\Phi^\nu(\mu)))$$ □

4.2 Examples

Example 4.1 The arbitrary functions $h, h' : \mathbf{B}^n \to \mathbf{B}^n$ fulfill the property that $(h, h')^{\smile} \in Hom^{\smile}(1_{\mathbf{B}^n}, 1_{\mathbf{B}^n})$. If h, h' are bijections, then $(h, h')^{\smile}$ is antiautomorphism.

Example 4.2 We take the functions $\Psi, h : \mathbf{B}^n \to \mathbf{B}^n$ arbitrary and $h' : \mathbf{B}^n \to \mathbf{B}^n$ defined by $\forall \mu \in \mathbf{B}^n, h'(\mu) = (0, \ldots, 0)$. We show that $(h, h')^{\smile} : 1_{\mathbf{B}^n} \to \Psi$ is an antimorphism: for any $\mu, \nu \in \mathbf{B}^n$ we infer

$$\Psi^{h'(\nu)}(h((1_{\mathbf{B}^n})^\nu(\mu))) = \Psi^{(0,\ldots,0)}(h(\mu)) = h(\mu).$$

Example 4.3 Let $\Phi : \mathbf{B}^2 \to \mathbf{B}^2$ the function from Figure 4.1: $\forall \mu \in \mathbf{B}^2, \forall \nu \in \mathbf{B}^2, \Phi^{\nu}(\mu_1, \mu_2) = (\mu_1, \overline{\nu_2}\mu_2 \cup \nu_2\overline{\mu_2})$ for which

$$\begin{aligned}
(\Phi^{\nu} \circ \Phi^{\nu})(\mu) &= \Phi^{\nu}(\mu_1, \overline{\nu_2}\mu_2 \cup \nu_2\overline{\mu_2}) \\
&= (\mu_1, \overline{\nu_2}(\overline{\nu_2}\mu_2 \cup \nu_2\overline{\mu_2}) \cup \nu_2\overline{\overline{\nu_2}\mu_2 \cup \nu_2\overline{\mu_2}}) \\
&= (\mu_1, \overline{\nu_2}\mu_2 \cup \nu_2(\nu_2 \cup \overline{\mu_2})(\overline{\nu_2} \cup \mu_2)) \\
&= (\mu_1, \overline{\nu_2}\mu_2 \cup \nu_2(\nu_2 \cup \overline{\mu_2})\mu_2) \\
&= (\mu_1, \overline{\nu_2}\mu_2 \cup \nu_2\mu_2) = (\mu_1, \mu_2)
\end{aligned}$$

is true. We have obtained that $(1_{\mathbf{B}^2}, 1_{\mathbf{B}^2})^{\frown} \in Aut^{\frown}(\Phi)$.

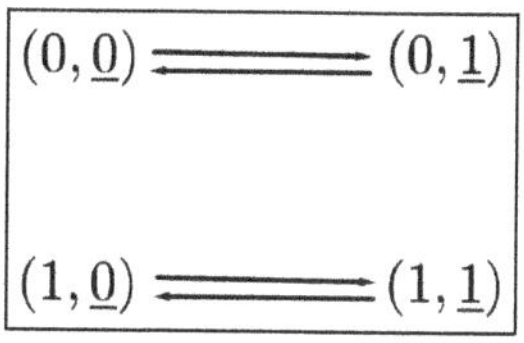

Figure 4.1 The function $\Phi(\mu_1, \mu_2) = (\mu_1, \overline{\mu_2})$.

Example 4.4 By following the previous idea, we show that the function Φ from Figure 4.1 and the function $\Psi : \mathbf{B}^2 \to \mathbf{B}^2$ from Figure 4.2 are antiisomorphic. Indeed, we have $\forall \mu \in \mathbf{B}^2, \Psi(\mu_1, \mu_2) = (\overline{\mu_1}, \mu_2)$ and $\forall \nu \in \mathbf{B}^2$,

$$\begin{aligned}
\Psi^{\nu}(\mu_1, \mu_2) &= (\overline{\nu_1}\mu_1 \cup \nu_1\overline{\mu_1}, \overline{\nu_2}\mu_2 \cup \nu_2\mu_2) \\
&= (\overline{\nu_1}\mu_1 \cup \nu_1\overline{\mu_1}, \mu_2).
\end{aligned}$$

We define $h, h' : \mathbf{B}^2 \to \mathbf{B}^2$ by $\forall \mu \in \mathbf{B}^2, h(\mu) = h'(\mu) = (\mu_2, \mu_1)$ and we infer:

$$\begin{aligned}
\Psi^{h'(\nu)}(h(\Phi^{\nu}(\mu))) &= \Psi^{(\nu_2,\nu_1)}(h(\mu_1, \overline{\nu_2}\mu_2 \cup \nu_2\overline{\mu_2})) \\
&= \Psi^{(\nu_2,\nu_1)}(\overline{\nu_2}\mu_2 \cup \nu_2\overline{\mu_2}, \mu_1) \\
&= (\overline{\nu_2}(\overline{\nu_2}\mu_2 \cup \nu_2\overline{\mu_2}) \cup \nu_2\Psi_1(\overline{\nu_2}\mu_2 \cup \nu_2\overline{\mu_2}, \mu_1), \\
&\qquad \overline{\nu_1}\mu_1 \cup \nu_1\Psi_2(\overline{\nu_2}\mu_2 \cup \nu_2\overline{\mu_2}, \mu_1)) \\
&= (\overline{\nu_2}\mu_2 \cup \nu_2\overline{\overline{\nu_2}\mu_2 \cup \nu_2\overline{\mu_2}}, \overline{\nu_1}\mu_1 \cup \nu_1\mu_1) \\
&= (\overline{\nu_2}\mu_2 \cup \nu_2(\nu_2 \cup \overline{\mu_2})(\overline{\nu_2} \cup \mu_2), \mu_1) \\
&= (\overline{\nu_2}\mu_2 \cup \nu_2(\nu_2 \cup \overline{\mu_2})\mu_2, \mu_1) = (\overline{\nu_2}\mu_2 \cup \nu_2\mu_2, \mu_1) \\
&= (\mu_2, \mu_1) = h(\mu),
\end{aligned}$$

thus $(h, h')^{\frown} \in Hom^{\frown}(\Phi, \Psi)$. But h, h' are bijections with $h = h^{-1}, h' = h'^{-1}$ and we can prove that $(h^{-1}, h'^{-1})^{\frown} \in Hom^{\frown}(\Psi, \Phi)$. We have $(h, h')^{\frown} \in Iso^{\frown}(\Phi, \Psi)$.

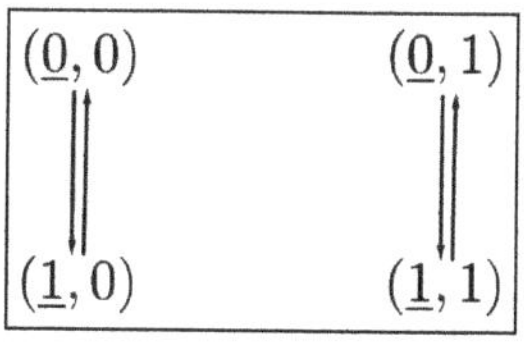

Figure 4.2 Function $\Psi(\mu_1, \mu_2) = (\overline{\mu_1}, \mu_2)$ which is antiisomorphic with the function $\Phi(\mu_1, \mu_2) = (\mu_1, \overline{\mu_2})$.

Example 4.5 Let $\Phi, \Psi : \mathbf{B}^n \to \mathbf{B}^n$ and $\mu' \in \mathbf{B}^n$ with the property that $\Psi(\mu') = \mu'$. We define $h : \mathbf{B}^n \to \mathbf{B}^n$ by $\forall \mu \in \mathbf{B}^n, h(\mu) = \mu'$ and we take $h' : \mathbf{B}^n \to \mathbf{B}^n$ arbitrary. Then $(h, h')^{\frown} \in Hom^{\frown}(\Phi, \Psi)$ because for any $\mu, \nu \in \mathbf{B}^n$ we have:

$$\Psi^{h'(\nu)}(h(\Phi^{\nu}(\mu))) = \Psi^{h'(\nu)}(\mu') = \mu' = h(\mu).$$

Example 4.6 We take h the constant function equal with μ', h' the constant function equal with $(0, \dots, 0) \in \mathbf{B}^n$ and Φ, Ψ arbitrary. Then $(h, h')^{\frown} \in Hom^{\frown}(\Phi, \Psi) : \forall \mu \in \mathbf{B}^n, \forall \nu \in \mathbf{B}^n$,

$$\Psi^{h'(\nu)}(h(\Phi^{\nu}(\mu))) = \Psi^{(0,\dots,0)}(\mu') = \mu' = h(\mu)$$

is true.

Example 4.7 We take $\Phi, \Psi : \mathbf{B}^n \to \mathbf{B}^n$ with $\forall \mu \in \mathbf{B}^n, \Phi_1(\mu) = \overline{\mu_1}, \Psi(\mu) = (\overline{\mu_1}, \overline{\mu_2}, \dots, \overline{\mu_n})$ and we define $h, h' : \mathbf{B}^n \to \mathbf{B}^n$ by $\forall \mu \in \mathbf{B}^n, h(\mu) = h'(\mu) = (\mu_1, \mu_1, \dots, \mu_1)$. For arbitrary $\mu, \nu \in \mathbf{B}^n$ we infer

$$\begin{aligned}\Psi^{h'(\nu)}(h(\Phi^{\nu}(\mu))) &= \Psi^{h'(\nu)}(h(\overline{\nu_1}\mu_1 \cup \nu_1\overline{\mu_1}, \Phi_2^{\nu_2}(\mu), \dots, \Phi_n^{\nu_n}(\mu))\\ &= \Psi^{(\nu_1,\nu_1,\dots,\nu_1)}(\overline{\nu_1}\mu_1 \cup \nu_1\overline{\mu_1}, \overline{\nu_1}\mu_1 \cup \nu_1\overline{\mu_1}, \dots, \overline{\nu_1}\mu_1 \cup \nu_1\overline{\mu_1}),\end{aligned}$$

and since

$$\begin{aligned}\overline{\nu_1}(\overline{\nu_1}\mu_1 \cup \nu_1\overline{\mu_1}) \cup \nu_1\overline{\overline{\nu_1}\mu_1 \cup \nu_1\overline{\mu_1}} &= \overline{\nu_1}\mu_1 \cup \nu_1(\nu_1 \cup \overline{\mu_1})(\overline{\nu_1} \cup \mu_1)\\ &= \overline{\nu_1}\mu_1 \cup \nu_1(\nu_1 \cup \overline{\mu_1})\mu_1\\ &= \overline{\nu_1}\mu_1 \cup \nu_1\mu_1 = \mu_1,\end{aligned}$$

we get

$$\Psi^{h'(\nu)}(h(\Phi^{\nu}(\mu))) = (\mu_1, \mu_1, \dots, \mu_1) = h(\mu),$$

in other words $(h, h')^{\frown} \in Hom^{\frown}(\Phi, \Psi)$.

Remark 4.4 In the previous examples, we note that $(h, h') \in Hom(\Phi, \Psi)$, $(h, h')^{\frown} \in Hom^{\frown}(\Phi, \Psi)$ are both true. For this, we can compare Example 3.2, page 37 with Example 4.2, Example 3.3, page 37 with Example 4.5, Example 3.4, page 37 with Example 4.6, and notice also that this happens for Examples 4.1, 4.3, 4.4, 4.7 too. The only way of trying to state a rule for this is contained in the unexpected question: do we have $Hom^{\frown}(\Phi, \Psi) \subset Hom(\Phi, \Psi)$?

4.3 The Composition

Theorem 4.2 Let the functions $\Omega, \Phi, \Psi, \Gamma : \mathbf{B}^n \longrightarrow \mathbf{B}^n$, together with the antimorphisms $(h, h')^{\frown} : \Phi \longrightarrow \Psi$, $(g, g')^{\frown} : \Psi \longrightarrow \Gamma$ and the morphisms $(f, f') : \Omega \longrightarrow \Phi$, $(i, i') : \Psi \longrightarrow \Gamma$.

(a) $(h \circ f, h' \circ f')^{\frown} : \Omega \longrightarrow \Psi, (i \circ h, i' \circ h')^{\frown} : \Phi \longrightarrow \Gamma$ are antimorphisms;
(b) $(g \circ h, g' \circ h') : \Phi \longrightarrow \Gamma$ is a morphism.

Proof: (a) We see that $\forall \nu \in \mathbf{B}^n$, the diagrams

$$\begin{array}{ccc} \mathbf{B}^n & \stackrel{\Omega^\nu}{\longrightarrow} & \mathbf{B}^n \\ f \downarrow & & \downarrow f \\ \mathbf{B}^n & \stackrel{\Phi^{f'(\nu)}}{\longrightarrow} & \mathbf{B}^n \\ h \downarrow & & \downarrow h \\ \mathbf{B}^n & \stackrel{\Psi^{(h' \circ f')(\nu)}}{\longleftarrow} & \mathbf{B}^n \end{array}$$

are all commutative, since $\forall \mu \in \mathbf{B}^n$,

$$\begin{aligned}(\Psi^{(h' \circ f')(\nu)} \circ (h \circ f) \circ \Omega^\nu)(\mu) &= (\Psi^{(h' \circ f')(\nu)} \circ h)((f \circ \Omega^\nu)(\mu)) \\ &= (\Psi^{(h' \circ f')(\nu)} \circ h)((\Phi^{f'(\nu)} \circ f)(\mu)) \\ &= ((\Psi^{(h' \circ f')(\nu)} \circ h \circ \Phi^{f'(\nu)}) \circ f)(\mu) \\ &= (h \circ f)(\mu)\end{aligned}$$

and similarly for $(i \circ h, i' \circ h')$.
(b) This is true because $\forall \nu \in \mathbf{B}^n$, the diagrams

$$\begin{array}{ccc} \mathbf{B}^n & \stackrel{\Phi^\nu}{\longrightarrow} & \mathbf{B}^n \\ h \downarrow & & \downarrow h \\ \mathbf{B}^n & \stackrel{\Psi^{h'(\nu)}}{\longleftarrow} & \mathbf{B}^n \\ g \downarrow & & \downarrow g \\ \mathbf{B}^n & \stackrel{\Gamma^{(g' \circ h')(\nu)}}{\longrightarrow} & \mathbf{B}^n \end{array}$$

commute: $\forall \mu \in \mathbf{B}^n$,

$$\begin{aligned}(\Gamma^{(g' \circ h')(\nu)} \circ (g \circ h))(\mu) &= ((\Gamma^{(g' \circ h')(\nu)} \circ g) \circ h)(\mu) = (\Gamma^{(g' \circ h')(\nu)} \circ g)(h(\mu)) \\ &= (\Gamma^{(g' \circ h')(\nu)} \circ g)((\Psi^{h'(\nu)} \circ h \circ \Phi^\nu)(\mu)) \\ &= ((\Gamma^{(g' \circ h')(\nu)} \circ g \circ \Psi^{h'(\nu)}) \circ (h \circ \Phi^\nu))(\mu) \\ &= (g \circ (h \circ \Phi^\nu))(\mu) = ((g \circ h) \circ \Phi^\nu)(\mu)\end{aligned}$$

□

Remark 4.5 It is not clear at this moment what happens if previously h, h' are bijections: is $(h^{-1}, h'^{-1})^{\frown} : \Psi \longrightarrow \Phi$ an antimorphism? This would mean that for any $\nu \in \mathbf{B}^n$, the commutativity of the diagram

$$\begin{array}{ccc} \mathbf{B}^n & \stackrel{\Phi^\nu}{\longrightarrow} & \mathbf{B}^n \\ h \downarrow & & \downarrow h \\ \mathbf{B}^n & \stackrel{\Psi^{h'(\nu)}}{\longleftarrow} & \mathbf{B}^n \end{array}$$

implies the commutativity of the diagram

$$\begin{array}{ccc} \mathbf{B}^n & \stackrel{\Phi^v}{\longrightarrow} & \mathbf{B}^n \\ h^{-1}\uparrow & & \uparrow h^{-1} \\ \mathbf{B}^n & \stackrel{\Psi^{h'(v)}}{\longleftarrow} & \mathbf{B}^n \end{array}$$

i.e. from

$$h = \Psi^{h'(v)} \circ h \circ \Phi^v$$

we get, since Φ^v and $\Psi^{h'(v)}$ are invertible:

$$\begin{aligned} h^{-1} &= (\Phi^v)^{-1} \circ h^{-1} \circ (\Psi^{h'(v)})^{-1} = \Phi^v \circ h^{-1} \circ \Psi^{h'(v)} \\ &= \Phi^{h'^{-1}(h'(v))} \circ h^{-1} \circ \Psi^{h'(v)}. \end{aligned} \tag{4.1}$$

In (4.1), the surjectivity of h' shows that $h'(v)$ runs through all the values of $\mathbf{B}^n$ when v runs through all the values of $\mathbf{B}^n$.

We cannot say when (4.1) holds. This would be the case if $\Phi^v = (\Phi^v)^{-1}$, $\Psi^{h'(v)} = (\Psi^{h'(v)})^{-1}$.

Definition 4.2 The previous antimorphisms $(h \circ f, h' \circ f')^\frown$, $(i \circ h, i' \circ h')^\frown$ are by definition the **composition** of the antimorphism $(h, h')^\frown$ with the morphism (f, f'), denoted $(h, h')^\frown \circ (f, f')$, and the **composition** of the morphism (i, i') with the antimorphism $(h, h')^\frown$, having the notation $(i, i') \circ (h, h')^\frown$.

Definition 4.3 The previous morphism $(g \circ h, g' \circ h')$ is by definition the **composition** of the antimorphisms $(g, g')^\frown$ and $(h, h')^\frown$ and it is usually denoted by $(g, g')^\frown \circ (h, h')^\frown$.

Remark 4.6 An asymmetry occurs by comparing Definition 3.1, page 35 with Definition 4.1, page 45 concerning the concepts of isomorphism and antiisomorphism: while $(h, h') \in Hom(\Phi, \Psi)$ and the bijectivity of h, h' imply $(h^{-1}, h'^{-1}) \in Hom(\Psi, \Phi)$, see Theorem 3.2, page 38, $(h, h')^\frown \in Hom^\frown(\Phi, \Psi)$ and the bijectivity of h, h' do not imply that $(h^{-1}, h'^{-1})^\frown \in Hom^\frown(\Psi, \Phi)$, see Remark 4.5.

Remark 4.7 When in Theorem 4.2 $(f, f'), (i, i')$ are isomorphisms and $(h, h')^\frown$ is an antiisomorphism, $(h \circ f, h' \circ f')^\frown$, $(i \circ h, i' \circ h')^\frown$ are antiisomorphisms (since the composition of bijections is a bijection). In particular, the composition of an automorphism with an antiautomorphism is an antiautomorphism.

Remark 4.8 If in Theorem 4.2, $(g, g')^\frown$ and $(h, h')^\frown$ are antiisomorphisms, then $(g \circ h, g' \circ h')$ is an isomorphism, in particular the composition of antiautomorphisms is an automorphism.

4.4 A Fixed Point Property

Theorem 4.3 The functions $\Phi, \Psi : \mathbf{B}^n \longrightarrow \mathbf{B}^n$ are given and $(h, h')^{\frown} : \Phi \longrightarrow \Psi$. If for $\mu \in \mathbf{B}^n$ we have $\Phi(\mu) = \mu$, then $\forall \nu \in \mathbf{B}^n, \Psi^{h'(\nu)}(h(\mu)) = h(\mu)$ and if h' is bijective, then we have that $\Psi(h(\mu)) = h(\mu)$.

Proof: For any $\nu \in \mathbf{B}^n$, we can write that

$$h(\mu) = (\Psi^{h'(\nu)} \circ h \circ \Phi^{\nu})(\mu) = \Psi^{h'(\nu)}(h(\Phi^{\nu}(\mu))) = \Psi^{h'(\nu)}(h(\mu)).$$

We suppose now that h' is bijective. Then $\nu' \in \mathbf{B}^n$ exists with $h'(\nu') = (1, \dots, 1) \in \mathbf{B}^n$ and we infer

$$h(\mu) = \Psi^{h'(\nu')}(h(\mu)) = \Psi^{(1,\dots,1)}(h(\mu)) = \Psi(h(\mu))$$ □

4.5 Antisymmetrical Functions Relative to Translations. Examples

Definition 4.4 The functions $\Phi, \Psi : \mathbf{B}^n \longrightarrow \mathbf{B}^n$ are called **antisymmetrical relative to the translation with** $\tau \in \mathbf{B}^n$, if $h' : \mathbf{B}^n \longrightarrow \mathbf{B}^n$ exists such that $(\theta^{\tau}, h')^{\frown} \in Iso^{\frown}(\Phi, \Psi)$, where $\forall \mu \in \mathbf{B}^n, \theta^{\tau}(\mu) = \mu \oplus \tau$.

Definition 4.5 We say that $\Phi : \mathbf{B}^n \longrightarrow \mathbf{B}^n$ is **antisymmetrical relative to the translation with** $\tau \in \mathbf{B}^n$, if $h' : \mathbf{B}^n \longrightarrow \mathbf{B}^n$ exists with the property that $(\theta^{\tau}, h')^{\frown} \in Aut^{\frown}(\Phi)$.

Example 4.8 We consider first the function $\Phi(\mu_1, \mu_2) = (\mu_1, \mu_1 \oplus \mu_2 \oplus 1)$ whose state portrait was drawn in Figure 4.3. We prove that it is antisymmetrical relative to the translation with $(0, 1)$: for any $\mu, \nu \in \mathbf{B}^2$ we can write

$$\begin{aligned}
(\Phi^{\nu} \circ \theta^{(0,1)} \circ \Phi^{\nu})(\mu) &= (\Phi^{\nu} \circ \theta^{(0,1)})(\mu_1, (\nu_2 \oplus 1)\mu_2 \oplus \nu_2(\mu_1 \oplus \mu_2 \oplus 1)) \\
&= \Phi^{\nu}(\mu_1, \mu_2 \oplus \mu_2\nu_2 \oplus \mu_1\nu_2 \oplus \mu_2\nu_2 \oplus \nu_2 \oplus 1) \\
&= \Phi^{\nu}(\mu_1, \mu_2 \oplus \mu_1\nu_2 \oplus \nu_2 \oplus 1) \\
&= (\mu_1, (\nu_2 \oplus 1)(\mu_2 \oplus \mu_1\nu_2 \oplus \nu_2 \oplus 1) \\
&\quad \oplus \nu_2(\mu_1 \oplus \mu_2 \oplus \mu_1\nu_2 \oplus \nu_2)) \\
&= (\mu_1, \mu_2\nu_2 \oplus \mu_1\nu_2 \oplus \nu_2 \oplus \nu_2 \oplus \mu_2 \oplus \mu_1\nu_2 \oplus \nu_2 \oplus 1 \\
&\quad \oplus \mu_1\nu_2 \oplus \mu_2\nu_2 \oplus \mu_1\nu_2 \oplus \nu_2) \\
&= (\mu_1, \mu_2 \oplus 1) = \theta^{(0,1)}(\mu).
\end{aligned}$$

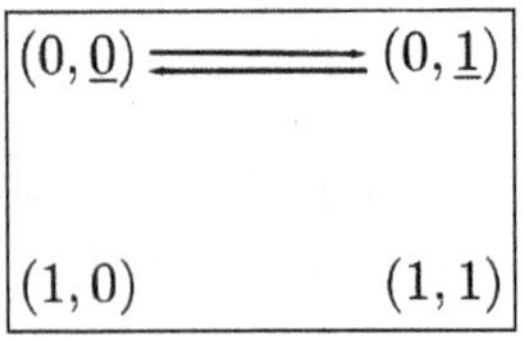

Figure 4.3 The function $\Phi(\mu_1, \mu_2) = (\mu_1, \mu_1 \oplus \mu_2 \oplus 1)$.

We can try to anticipate the fact that $(\theta^{(0,1)}, 1_{\mathbf{B}^2})^{\smallfrown} \in Aut^{\smallfrown}(\Phi)$ in the following way: if we add (0, 1) in Figure 4.3 to all the points of $\mathbf{B}^2$ we get Figure 4.3 again. So far, we have that $(\theta^{(0,1)}, 1_{\mathbf{B}^2}) \in Aut(\Phi)$. But $\forall \nu \in \mathbf{B}^2, \Phi^\nu = (\Phi^\nu)^{-1}$, thus the property follows.

Example 4.9 In order to construct a new function Ψ that is antisymmetrical with Φ from Figure 4.3, we add in that figure $(1, 0)$ to all the points $\mu \in \mathbf{B}^2$; we obtain Figure 4.4 and $\Psi(\mu_1, \mu_2) = (\mu_1, \mu_1 \oplus \mu_2)$. We prove that $(\theta^{(1,0)}, 1_{\mathbf{B}^2})^{\smallfrown} \in Iso^{\smallfrown}(\Phi, \Psi)$: for any $\mu, \nu \in \mathbf{B}^2$, we infer

$$\begin{aligned}
(\Psi^\nu \circ \theta^{(1,0)} \circ \Phi^\nu)(\mu) &= (\Psi^\nu \circ \theta^{(1,0)})(\mu_1, (\nu_2 \oplus 1)\mu_2 \oplus \nu_2(\mu_1 \oplus \mu_2 \oplus 1)) \\
&= \Psi^\nu(\mu_1 \oplus 1, \mu_2 \oplus \mu_2\nu_2 \oplus \mu_1\nu_2 \oplus \mu_2\nu_2 \oplus \nu_2) \\
&= \Psi^\nu(\mu_1 \oplus 1, \mu_2 \oplus \mu_1\nu_2 \oplus \nu_2) \\
&= (\mu_1 \oplus 1, (\nu_2 \oplus 1)(\mu_2 \oplus \mu_1\nu_2 \oplus \nu_2) \\
&\quad \oplus \nu_2(\mu_1 \oplus 1 \oplus \mu_2 \oplus \mu_1\nu_2 \oplus \nu_2)) \\
&= (\mu_1 \oplus 1, \mu_2\nu_2 \oplus \mu_1\nu_2 \oplus \nu_2 \oplus \mu_2 \oplus \mu_1\nu_2 \oplus \nu_2 \\
&\quad \oplus \mu_1\nu_2 \oplus \nu_2 \oplus \mu_2\nu_2 \oplus \mu_1\nu_2 \oplus \nu_2) \\
&= (\mu_1 \oplus 1, \mu_2) = \theta^{(1,0)}(\mu).
\end{aligned}$$

Once again, by adding $(1, 0)$ to all the points $\mu \in \mathbf{B}^2$ in Figure 4.3 we have obtained Ψ from Figure 4.4 with $(\theta^{(1,0)}, 1_{\mathbf{B}^2}) \in Iso(\Phi, \Psi)$. It happens that Ψ fulfills $\forall \nu \in \mathbf{B}^2, \Psi^\nu = (\Psi^\nu)^{-1}$, wherefrom $(\theta^{(1,0)}, 1_{\mathbf{B}^2})^{\smallfrown} \in Iso^{\smallfrown}(\Phi, \Psi)$.

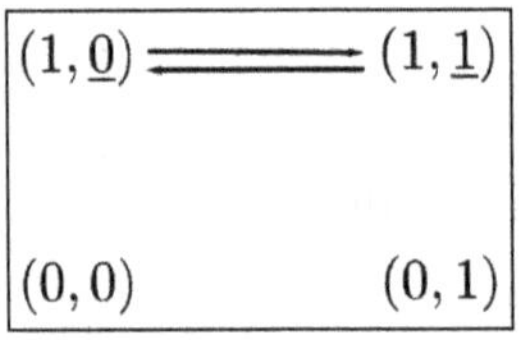

Figure 4.4 The function $\Psi(\mu_1, \mu_2) = (\mu_1, \mu_1 \oplus \mu_2)$.

4.6 Antimorphisms vs Predecessors and Successors

Theorem 4.4 The functions $\Phi, \Psi : \mathbf{B}^n \to \mathbf{B}^n$ are given, together with $\mu \in \mathbf{B}^n$ and the antimorphism $(h, h')^{\smallfrown} \in Hom^{\smallfrown}(\Phi, \Psi)$.

(a) We have:

$$h(\mu_{\Phi}^{-}) \subset h(\mu)_{\Psi}^{+}, \tag{4.2}$$

$$h(\mu_{\Phi}^{+}) \subset h(\mu)_{\Psi}^{-}, \tag{4.3}$$

$$h(O_{\Phi}^{-}(\mu)) \subset O_{\Psi}^{+}(h(\mu)), \tag{4.4}$$

$$h(O_{\Phi}^{+}(\mu)) \subset O_{\Psi}^{-}(h(\mu)). \tag{4.5}$$

(b) If $(h, h')^{\frown} \in Iso^{\frown}(\Phi, \Psi)$, then

$$h(\mu_{\Phi}^{-}) = h(\mu)_{\Psi}^{+}, \tag{4.6}$$

$$h(\mu_{\Phi}^{+}) = h(\mu)_{\Psi}^{-}, \tag{4.7}$$

$$h(O_{\Phi}^{-}(\mu)) = O_{\Psi}^{+}(h(\mu)), \tag{4.8}$$

$$h(O_{\Phi}^{+}(\mu)) = O_{\Psi}^{-}(h(\mu)). \tag{4.9}$$

Proof: (a) (4.2): We take $\lambda \in \mu_{\Phi}^{-}$ arbitrary, thus $\nu \in \mathbf{B}^n$ exists with $\mu = \Phi^{\nu}(\lambda)$. As

$$h(\lambda) = \Psi^{h'(\nu)}(h(\Phi^{\nu}(\lambda))) = \Psi^{h'(\nu)}(h(\mu)),$$

we infer $h(\lambda) \in h(\mu)_{\Psi}^{+}$.

(4.5): Let $\lambda \in O_{\Phi}^{+}(\mu)$ arbitrary, i.e. $\nu \in \mathbf{B}^n, \dots, \nu' \in \mathbf{B}^n$ exist such that $\lambda = (\Phi^{\nu} \circ \dots \circ \Phi^{\nu'})(\mu)$. From

$$\begin{aligned} h(\mu) &= (\Psi^{h'(\nu')} \circ h \circ \Phi^{\nu'})(\mu) \\ &= \dots = (\Psi^{h'(\nu')} \circ \dots \circ \Psi^{h'(\nu)} \circ h \circ \Phi^{\nu} \circ \dots \circ \Phi^{\nu'})(\mu) \\ &= (\Psi^{h'(\nu')} \circ \dots \circ \Psi^{h'(\nu)})(h((\Phi^{\nu} \circ \dots \circ \Phi^{\nu'})(\mu))) \\ &= (\Psi^{h'(\nu')} \circ \dots \circ \Psi^{h'(\nu)})(h(\lambda)), \end{aligned}$$

we get $h(\lambda) \in O_{\Psi}^{-}(h(\mu))$.

(b) We notice first of all that the bijectivity of h, h' implies that $\forall \nu \in \mathbf{B}^n, \Phi^{\nu}, \Psi^{\nu}$ are bijective.

(4.6): We prove

$$h(\mu)_{\Psi}^{+} \subset h(\mu_{\Phi}^{-}). \tag{4.10}$$

Let $\lambda' \in h(\mu)_{\Psi}^{+}$ arbitrary, thus $\nu' \in \mathbf{B}^n$ exists such that $\lambda' = \Psi^{\nu'}(h(\mu))$. The bijectivity of h' shows the existence of $\nu \in \mathbf{B}^n$ with $\nu' = h'(\nu)$. In addition, the bijectivity of Φ^{ν} gives the existence of $\lambda \in \mathbf{B}^n$ with $\mu = \Phi^{\nu}(\lambda)$, wherefrom $\lambda \in \mu_{\Phi}^{-}$. We have

$$\lambda' = \Psi^{\nu'}(h(\mu)) = \Psi^{h'(\nu)}(h(\Phi^{\nu}(\lambda))) = h(\lambda),$$

where $h(\lambda) \in h(\mu_{\Phi}^{-})$. The inclusion (4.10) is proved and, if we take into account (4.2) also, (4.6) results.

(4.9): We prove

$$O^-_{\Psi}(h(\mu)) \subset h(O^+_{\Phi}(\mu)) \tag{4.11}$$

and let $\lambda' \in O^-_{\Psi}(h(\mu))$ arbitrary. We have the existence of $\nu' \in \mathbf{B}^n, \dots, \delta' \in \mathbf{B}^n$ with $h(\mu) = (\Psi^{\nu'} \circ \dots \circ \Psi^{\delta'})(\lambda')$. We define $\nu, \dots, \delta, \lambda \in \mathbf{B}^n$ by $\nu' = h'(\nu), \dots, \delta' = h'(\delta), \lambda' = h(\lambda)$ and we can write:

$$\begin{aligned}
(\Psi^{\nu'} \circ \dots \circ \Psi^{\delta'})(\lambda') &= h(\mu) = (\Psi^{h'(\nu)} \circ h \circ \Phi^{\nu})(\mu) = \dots \\
&= (\Psi^{h'(\nu)} \circ \dots \circ \Psi^{h'(\delta)} \circ h \circ \Phi^{\delta} \circ \dots \circ \Phi^{\nu})(\mu) \\
&= (\Psi^{h'(\nu)} \circ \dots \circ \Psi^{h'(\delta)})(h((\Phi^{\delta} \circ \dots \circ \Phi^{\nu})(\mu))) \\
&= (\Psi^{\nu'} \circ \dots \circ \Psi^{\delta'})(h((\Phi^{\delta} \circ \dots \circ \Phi^{\nu})(\mu)))
\end{aligned}$$

thus

$$h(\lambda) = \lambda' = h((\Phi^{\delta} \circ \dots \circ \Phi^{\nu})(\mu)),$$

$\lambda = (\Phi^{\delta} \circ \dots \circ \Phi^{\nu})(\mu) \in O^+_{\Phi}(\mu)$ (since $\Psi^{\nu'}, \dots, \Psi^{\delta'}$ and h are bijections), and we infer that $\lambda' \in h(O^+_{\Phi}(\mu))$. The inclusion (4.11) is proved and, if we take into account (4.5), Eq. (4.9) follows □

Corollary 4.1 We suppose that $(h, h')^{\smallfrown} \in Iso^{\smallfrown}(\Phi, \Psi)$ and let $\mu \in \mathbf{B}^n$.

(a) μ is a source of Φ if and only if $h(\mu)$ is a sink of Ψ;

(b) μ is an isolated fixed point of Φ if and only if $h(\mu)$ is an isolated fixed point of Ψ;

(c) μ is a transient point of Φ if and only if $h(\mu)$ is a transient point of Ψ;

(d) μ is a sink of Φ if and only if $h(\mu)$ is a source of Ψ.

Proof: The function h is bijective and we use Eqs. (4.6) and (4.7) that give:

$$\begin{aligned}
&card(\mu^-_{\Phi}) = card(h(\mu^-_{\Phi})) = card(h(\mu)^+_{\Psi}), \\
&card(\mu^+_{\Phi}) = card(h(\mu^+_{\Phi})) = card(h(\mu)^-_{\Psi})
\end{aligned}$$

□

5

Invariant Sets

In the dynamical systems theory (synchronicity, real time, real space), the set A is (positively) invariant with respect to $\dot{x} = \Phi(x)$ [14, 17] if $x(0) \in A \Longrightarrow \forall t \geq 0, x(t) \in A$; alternatively (synchronicity, discrete time, real space) the set A is [3, 11] (positively, or forward) invariant relatively to $x_{k+1} = \Phi(x_k)$ if $x_0 \in A \Longrightarrow \forall k \geq 0, x_k \in A$, in other words if $\Phi(A) \subset A$.

A different point of view on this concept gives what we might call strong invariance. The set $A \subset X$ is [14], see also [1], invariant (synchronicity, real time, real space) if $\forall t \geq 0, \Phi_t(A) = A$, where $\Phi_t : X \longrightarrow X$ is a flow, $t \in \mathbf{R}$. In the same interpretation of invariance we have that A is [11] invariant (synchronicity, discrete time, real space) if $\Phi(A) = A$.

Therefore, we have formalized in two versions the fact that the set $A \subset \mathbf{B}^n$ is invariant, i.e.that Φ brings the elements of A in A.

The intersection and the union of the invariant sets is invariant. The sets of fixed points of Φ are invariant.

The morphisms and the antimorphisms bring invariant sets in invariant sets, in particular if a function is symmetrical or antisymmetrical relative to a translation, then the translation of an invariant set is invariant.

The sets $A_1, \ldots, A_k \subset \mathbf{B}^n$ are called relatively isolated if they are invariant and disjoint two by two. The isomorphisms and the antiisomorphisms bring relatively isolated sets in relatively isolated sets.

5.1 Definition

Lemma 5.1 We consider the function $\Psi : A \longrightarrow A$, where $A \subset \mathbf{B}^n, A \neq \emptyset$. The following statements are equivalent:

(a) Ψ is bijective;
(b) Ψ is injective;
(c) Ψ is surjective.

Boolean Functions: Topics in Asynchronicity, First Edition. Serban E. Vlad.

Proof: (a) $\Longrightarrow$ (b) Obvious.

(b) $\Longrightarrow$ (c) We suppose that Ψ is not surjective, i.e. $\exists \mu^* \in A, \forall \mu \in A, \Psi(\mu) \neq \mu^*$, hence $\Psi(A) \subset A, \Psi(A) \neq A$. In this situation, we have $card(\Psi(A)) < card(A)$ and $\mu, \mu' \in A$ exist with the property that $\mu \neq \mu'$ and $\Psi(\mu) = \Psi(\mu')$, therefore Ψ is not injective, contradiction.

(c) $\Longrightarrow$ (a) Let us suppose against all reason that this is false, meaning that Ψ is not injective. We infer the existence of $\mu, \mu' \in A$ with $\mu \neq \mu'$ and $\Psi(\mu) = \Psi(\mu')$. This implies that $\Psi(A) \subset A, \Psi(A) \neq A$, thus Ψ is not surjective, contradiction □

Definition 5.1 Let the function $\Phi : \mathbf{B}^n \longrightarrow \mathbf{B}^n$ and the set $A \subset \mathbf{B}^n, A \neq \emptyset$. The relations

$$\forall \nu \in \mathbf{B}^n, \Phi^\nu(A) \subset A, \tag{5.1}$$

$$\forall \nu \in \mathbf{B}^n, \Phi^\nu(A) = A \tag{5.2}$$

are called of **invariance** of A. We say that A is a k-**invariant set,** $k \in \{(5.1), (5.2)\}$ (relative to Φ).

Remark 5.1 Obviously the implication (5.2)$\Longrightarrow$(5.1) holds and we can think of (5.2) as of a strong version of (5.1), where for all $\nu \in \mathbf{B}^n$, the function $A \ni \mu \mapsto \Phi^\nu(\mu) \in A$ is surjective. Lemma 5.1 shows that surjectivity is equivalent with bijectivity.

Remark 5.2 The (5.1)-invariance of $\mathbf{B}^n$ is true and this brings a triviality here. We have avoided in Definition 5.1, for convenience, the other triviality: $A = \emptyset$.

Theorem 5.1 For $\emptyset \subsetneq A \subset \mathbf{B}^n$ and $\overline{A} = \{\overline{\mu} | \mu \in A\}$, we have the equivalencies:

$$\forall \nu \in \mathbf{B}^n, \Phi^\nu(A) \subset A \Longleftrightarrow \forall \nu \in \mathbf{B}^n, \Phi^{*\nu}(\overline{A}) \subset \overline{A}, \tag{5.3}$$

$$\forall \nu \in \mathbf{B}^n, \Phi^\nu(A) = A \Longleftrightarrow \forall \nu \in \mathbf{B}^n, \Phi^{*\nu}(\overline{A}) = \overline{A}. \tag{5.4}$$

Proof: (5.3) $\Longrightarrow$ Let $\nu \in \mathbf{B}^n, \delta \in \overline{A}$ arbitrary and we denote $\mu = \overline{\delta}$. We see that $\mu \in A$ and

$$\Phi^{*\nu}(\delta) = \Phi^{\nu *}(\delta) = \overline{\Phi^\nu(\overline{\delta})} = \overline{\Phi^\nu(\mu)} \in \overline{A}.$$

(5.4) $\Longleftarrow$ Let $\nu \in \mathbf{B}^n, \mu' \in A$ arbitrary and we use the notation $\delta' = \overline{\mu'}$. Then $\delta' \in \overline{A}$ and the hypothesis states the existence of $\delta \in \overline{A}$ with $\Phi^{*\nu}(\delta) = \delta'$. We denote $\mu = \overline{\delta}$, thus $\mu \in A$. We infer that

$$\Phi^\nu(\mu) = \overline{\Phi^{\nu *}(\overline{\mu})} = \overline{\Phi^{*\nu}(\delta)} = \overline{\delta'} = \mu'.$$

The surjectivity of Φ^ν follows □

5.2 Examples

Example 5.1 The identity $1_{\mathbf{B}^n} : \mathbf{B}^n \longrightarrow \mathbf{B}^n$ satisfies that any $A \subset \mathbf{B}^n, A \neq \emptyset$ is (5.1)-invariant and (5.2)-invariant.

Example 5.2 In Figure 5.1, the only (5.1)-invariant set $A \subset \mathbf{B}^2$ is $\mathbf{B}^2$ itself. On the other hand, the (5.2)-invariance of $\mathbf{B}^2$ does not hold since, for example, $\Phi^{(1,1)}(\mathbf{B}^2) = \mathbf{B}^2$ and $\Phi^{(1,0)}(\mathbf{B}^2) = \{(1,0),(0,1)\}$.

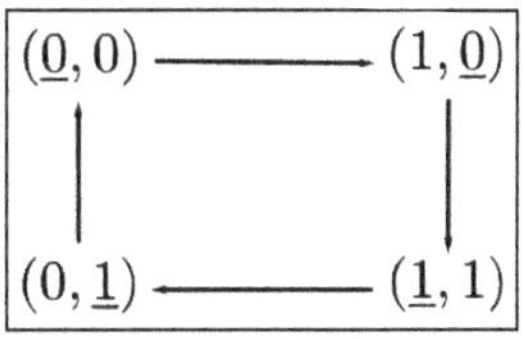

Figure 5.1 $\mathbf{B}^2$ is (5.1)-invariant.

Example 5.3 We consider Figure 5.2, where the sets $\{(0,0)\}, \{(0,1)\}$ and $\{(0,0),\ (0,1)\}$ are (5.1)-invariant and (5.2)-invariant. Note that these sets contain the fixed points of Φ. The sets $\{(1,0),(1,1),(0,1)\}$ and $\{(1,1),(0,1)\}$ from the same figure are (5.1)-invariant only.

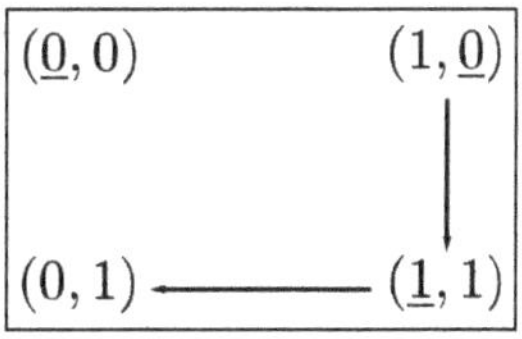

Figure 5.2 The sets $\{(0,0)\}, \{(0,1)\}, \{(0,0),(0,1)\}$ are (5.1)-invariant and (5.2)-invariant; $\{(1,0),(1,1),(0,1)\}$, $\{(1,1),(0,1)\}$ are (5.1)-invariant.

Example 5.4 The sets $\{(0,0),(0,1)\}, \{(1,0),(1,1)\}$ and $\mathbf{B}^2$ from Figure 5.3 are (5.1)-invariant and (5.2)-invariant. There are no fixed points of Φ associated with the (5.2)-invariance of the three sets in this case.

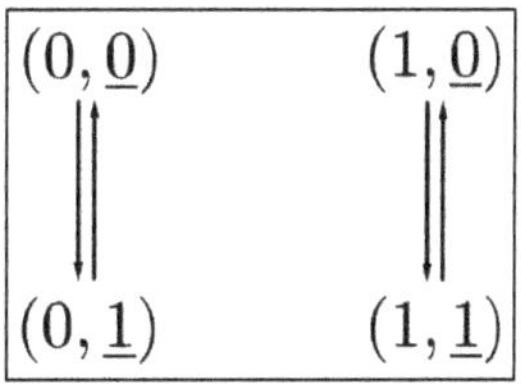

Figure 5.3 The sets $\{(0,0),(0,1)\}, \{(1,0),(1,1)\}$ and $\mathbf{B}^2$ are (5.1)-invariant and (5.2)-invariant.

Remark 5.3 The previous example and many other examples of functions that fulfill (5.2) suggest that the satisfaction of (5.2) might imply

$$\forall \mu \in A, \forall \nu \in \mathbf{B}^n, (\Phi^\nu \circ \Phi^\nu)(\mu) = \mu, \tag{5.5}$$

see Remark 4.5, page 49. This implication is false, for example the function $\Phi : \mathbf{B}^2 \longrightarrow \mathbf{B}^2$ defined by $\forall \mu \in \mathbf{B}^2$,

$$\Phi(\mu) = (\overline{\mu_1}, \mu_1 \oplus \mu_2)$$

satisfies (5.2) for $A = \mathbf{B}^2$:

$$\forall \nu \in \mathbf{B}^n, \Phi^\nu(\mathbf{B}^2) = \mathbf{B}^2,$$

but $\Phi \circ \Phi \neq 1_{\mathbf{B}^2}$, i.e. (5.5) is false for $A = \mathbf{B}^2$.

5.3 Properties

Theorem 5.2 Let $A \subset \mathbf{B}^n, A \neq \emptyset$ and the function $\Phi : \mathbf{B}^n \longrightarrow \mathbf{B}^n$. Then the (5.1)-invariance property of A is equivalent with

$$\forall \mu \in A, \mu^+ \subset A. \tag{5.6}$$

Proof: (5.1) $\Longrightarrow$ (5.6) Let $\mu \in A$ and $\lambda \in \mu^+$ arbitrary, thus $\nu \in \mathbf{B}^n$ exists such that $\lambda = \Phi^\nu(\mu)$. The hypothesis states that $\lambda \in A$.

(5.6) $\Longrightarrow$ (5.1) We take $\mu \in A$ and $\nu \in \mathbf{B}^n$ arbitrary. We have $\Phi^\nu(\mu) \in \mu^+$, thus $\Phi^\nu(\mu) \in A$ □

Theorem 5.3 For any $\mu \in \mathbf{B}^n$, and any $k \in \{(5.1), (5.2)\}$, $A = \{\mu\}$ is k-invariant $\Longleftrightarrow \Phi(\mu) = \mu$.

Proof: We choose to refer to the (5.1)-invariance.

$\Longrightarrow \forall \nu \in \mathbf{B}^n, \Phi^\nu(\{\mu\}) \subset \{\mu\}$ implies $\forall \nu \in \mathbf{B}^n, \Phi^\nu(\mu) = \mu$, i.e. for $\nu = (1, \dots, 1)$ we obtain $\Phi(\mu) = \mu$.

$\Longleftarrow$ From $\Phi(\mu) = \mu$ we infer $\forall \nu \in \mathbf{B}^n, \Phi^\nu(\mu) = \mu$ thus $\forall \nu \in \mathbf{B}^n$, $\Phi^\nu(\{\mu\}) \subset \{\mu\}$ □

Theorem 5.4 If $A = \{\mu^1, \dots, \mu^p\} \subset \mathbf{B}^n, p \geq 2$ then A is (5.1)-invariant if and only if $\{\mu^1, \dots, \mu^p\} = \mu^{1+} \cup \dots \cup \mu^{p+}$.

Proof: If. We take $i \in \{1, \dots, p\}$ and $\nu \in \mathbf{B}^n$ arbitrary, thus $\Phi^\nu(\mu^i) \in \mu^{i+}$. We infer $\Phi^\nu(\mu^i) \in \{\mu^1, \dots, \mu^p\}$.

Only if. $\{\mu^1, \dots, \mu^p\} \subset \mu^{1+} \cup \dots \cup \mu^{p+}$ As $\{\mu^1\} \subset \mu^{1+}, \dots, \{\mu^p\} \subset \mu^{p+}$ we infer $\{\mu^1, \dots, \mu^p\} = \{\mu^1\} \cup \dots \cup \{\mu^p\} \subset \mu^{1+} \cup \dots \cup \mu^{p+}$.

$\mu^{1+} \cup \ldots \cup \mu^{p+} \subset \{\mu^1, \ldots, \mu^p\}$ Let $i \in \{1, \ldots, p\}$ arbitrary. We know from Theorem 5.2 that $\mu^{i+} \subset \{\mu^1, \ldots, \mu^p\}$, thus the inclusion follows □

Theorem 5.5 If for $A \subset \mathbf{B}^n, A \neq \emptyset$ we have $\forall \mu \in A, \Phi(\mu) = \mu$, then (5.2) is true.

Proof: Let $\nu \in \mathbf{B}^n$ arbitrary. For any $\mu \in A$ we have $\Phi^\nu(\mu) = \Phi(\mu) = \mu$. This shows that (5.1) is true and, on the other hand, that the $A \longrightarrow A$ function $A \ni \mu \mapsto \Phi^\nu(\mu) \in A$ is injective. The surjectivity expressed by statement (5.2) holds from Lemma 5.1, page 55 □

Theorem 5.6 Let $A, B \subset \mathbf{B}^n, A \neq \emptyset, B \neq \emptyset$.

(a) We suppose that A and B are (5.1)-invariant; then[1] $A \cup B$ is (5.1)-invariant and if $A \cap B \neq \emptyset$, then $A \cap B$ is (5.1)-invariant.

(b) If A and B are (5.2)-invariant, then $A \cup B$ is (5.2)-invariant and if, in addition, $A \cap B \neq \emptyset$, then $A \cap B$ is (5.2)-invariant.

Proof: (a) We take $\mu \in A \cup B$ and $\nu \in \mathbf{B}^n$ arbitrary. If $\mu \in A$, then $\Phi^\nu(A) \in A \subset A \cup B$ due to the (5.1)-invariance of A and similarly for $\mu \in B$. $A \cup B$ is (5.1)-invariant.

We suppose now that $A \cap B \neq \emptyset$ and we take $\mu \in A \cap B$ and $\nu \in \mathbf{B}^n$ arbitrary. $\Phi^\nu(\mu) \in A, \Phi^\nu(\mu) \in B$ result from the (5.1)-invariance of A, B, hence $\Phi^\nu(\mu) \in A \cap B$. We have obtained that $A \cap B$ is (5.1)-invariant.

(b) We take $\nu \in \mathbf{B}^n$ arbitrary; $\Phi^\nu(A \cup B) \subset A \cup B$ follows from (a) and on the other hand the surjectivities of $\Psi : A \longrightarrow A, \Lambda : B \longrightarrow B$ which are defined by

$$\forall \mu \in A, \Psi(\mu) = \Phi^\nu(\mu),$$

$$\forall \mu \in B, \Lambda(\mu) = \Phi^\nu(\mu)$$

imply the surjectivity of $\Omega : A \cup B \longrightarrow A \cup B$,

$$\forall \mu \in A \cup B, \Omega(\mu) = \Phi^\nu(\mu),$$

showing the (5.2)-invariance of $A \cup B$. In addition, if $A \cap B \neq \emptyset$, the inclusion $\Phi^\nu(A \cap B) \subset A \cap B$ follows from (a) and the injectivity of Ψ implies the injectivity of $\Gamma : A \cap B \longrightarrow A \cap B$,

$$\forall \mu \in A \cap B, \Gamma(\mu) = \Phi^\nu(\mu).$$

This proves that $A \cap B$ is (5.2)-invariant □

---

1 One notation "∪" for two different laws occurs in this book: the union of the binary numbers and the union of the sets, but no possibility of confusion exists.

Remark 5.4 Two properties that are equivalent with the (5.1)-invariance of $A = \{\mu^1, \dots, \mu^p\}$ are $\forall \mu \in A, O^+(\mu) \subset A$, see Theorem 5.2 and $\{\mu^1, \dots, \mu^p\} = O^+(\mu^1) \cup \dots \cup O^+(\mu^p)$, see Theorem 5.4.

5.4 Homomorphic Functions vs Invariant Sets

Theorem 5.7 The functions $\Phi, \Psi : \mathbf{B}^n \longrightarrow \mathbf{B}^n$ and the set $A \subset \mathbf{B}^n, A \neq \emptyset$ are given. We suppose that

$$\forall \nu \in \mathbf{B}^n, \Phi^\nu(A) \subset A \tag{5.7}$$

holds and let $(h, h') \in Hom(\Phi, \Psi)$.

(a) We have

$$\forall \nu \in \mathbf{B}^n, \Psi^{h'(\nu)}(h(A)) \subset h(A); \tag{5.8}$$

(b) if h' is bijective, then the invariance

$$\forall \nu \in \mathbf{B}^n, \Psi^\nu(h(A)) \subset h(A) \tag{5.9}$$

holds.

Proof: (a) We take $\mu \in h(A)$ and $\nu \in \mathbf{B}^n$ arbitrary, thus $\lambda \in A$ exists such that $h(\lambda) = \mu$. We infer $\Phi^\nu(\lambda) \in A$, thus $h(\Phi^\nu(\lambda)) \in h(A)$. But $\Psi^{h'(\nu)}(\mu) = h(\Phi^\nu(\lambda))$, wherefrom $\Psi^{h'(\nu)}(\mu) \in h(A)$.

(b) Obvious, since (a) holds in this case with h' surjective □

Theorem 5.8 We suppose that

$$\forall \nu \in \mathbf{B}^n, \Phi^\nu(A) = A \tag{5.10}$$

and let $(h, h') \in Iso(\Phi, \Psi)$. Then

$$\forall \nu \in \mathbf{B}^n, \Psi^\nu(h(A)) = h(A). \tag{5.11}$$

Proof: Let $\mu, \mu' \in h(A), \mu \neq \mu'$ for which $\lambda, \lambda' \in A$ exist such that $h(\lambda) = \mu, h(\lambda') = \mu'$ and, because h is bijective, we get $\lambda \neq \lambda'$. We take $\nu \in \mathbf{B}^n$ arbitrarily. As $\Phi^\nu(A) = A$ we infer $\Phi^\nu(\lambda) \neq \Phi^\nu(\lambda')$ and therefore $h(\Phi^\nu(\lambda)) \neq h(\Phi^\nu(\lambda'))$. We obtain

$$\begin{aligned}\Psi^{h'(\nu)}(\mu) &= \Psi^{h'(\nu)}(h(\lambda)) = h(\Phi^\nu(\lambda)) \\ &\neq h(\Phi^\nu(\lambda')) = \Psi^{h'(\nu)}(h(\lambda')) = \Psi^{h'(\nu)}(\mu').\end{aligned}$$

As the inclusion $\Psi^{h'(\nu)}(h(A)) \subset h(A)$ is true from Theorem 5.7, we get that the function $\Omega : h(A) \longrightarrow h(A)$ defined by

$$\forall \mu \in h(A), \Omega(\mu) = \Psi^{h'(\nu)}(\mu)$$

is injective, therefore it is also surjective from Lemma 5.1, page 55. The statement of the theorem follows from the fact that h' is bijective □

Example 5.5 Let the functions $\Phi, \Psi, h, h' : \mathbf{B}^2 \longrightarrow \mathbf{B}^2$, given by $\forall \mu \in \mathbf{B}^2$,

$$\Phi(\mu) = (\mu_1 \cup \mu_2, \overline{\mu_1} \cup \mu_2),$$
$$\Psi(\mu) = (\mu_1, \overline{\mu_1} \cup \mu_2),$$
$$h(\mu) = (\mu_1\overline{\mu_2}, \mu_2),$$
$$h'(\mu) = (\mu_1, \mu_2).$$

The state portraits of Φ and Ψ were drawn in Figures 5.4 and 5.5.

We compute for any $\mu, \nu \in \mathbf{B}^2$,

$$\begin{aligned}
(h \circ \Phi^{\nu})(\mu) &= h(\overline{\nu_1}\mu_1 \cup \nu_1(\mu_1 \cup \mu_2), \overline{\nu_2}\mu_2 \cup \nu_2(\overline{\mu_1} \cup \mu_2)) \\
&= ((\overline{\nu_1}\mu_1 \cup \nu_1(\mu_1 \cup \mu_2))\overline{\overline{\nu_2}\mu_2 \cup \nu_2(\overline{\mu_1} \cup \mu_2)}, \\
&\quad \overline{\nu_2}\mu_2 \cup \nu_2(\overline{\mu_1} \cup \mu_2)) \\
&= ((\overline{\nu_1}\mu_1 \cup \nu_1\mu_1 \cup \nu_1\mu_2)(\nu_2 \cup \overline{\mu_2})(\overline{\nu_2} \cup \mu_1\overline{\mu_2}), \\
&\quad \overline{\nu_2}\mu_2 \cup \nu_2\overline{\mu_1} \cup \nu_2\mu_2) \\
&= ((\mu_1 \cup \nu_1\mu_2)(\nu_2\mu_1\overline{\mu_2} \cup \overline{\nu_2}\,\overline{\mu_2} \cup \mu_1\overline{\mu_2}), \mu_2 \cup \nu_2\overline{\mu_1}) \\
&= (\mu_1(\nu_2\mu_1\overline{\mu_2} \cup \overline{\nu_2}\,\overline{\mu_2} \cup \mu_1\overline{\mu_2}), \mu_2 \cup \nu_2\overline{\mu_1}) \\
&= (\nu_2\mu_1\overline{\mu_2} \cup \overline{\nu_2}\mu_1\overline{\mu_2} \cup \mu_1\overline{\mu_2}, \mu_2 \cup \nu_2\overline{\mu_1}) \\
&= (\mu_1\overline{\mu_2} \cup \mu_1\overline{\mu_2}, \mu_2 \cup \nu_2\overline{\mu_1}) = (\mu_1\overline{\mu_2}, \mu_2 \cup \nu_2\overline{\mu_1})
\end{aligned}$$

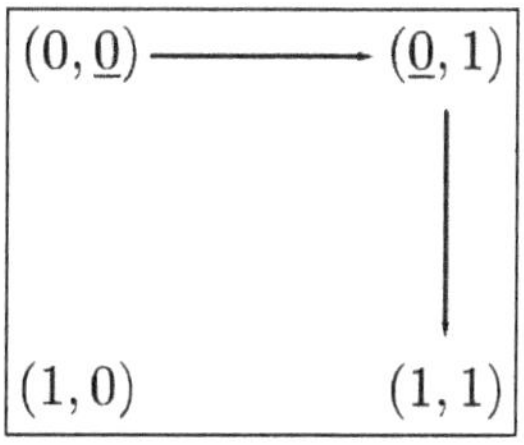

Figure 5.4 The function $\Phi(\mu) = (\mu_1 \cup \mu_2, \overline{\mu_1} \cup \mu_2)$.

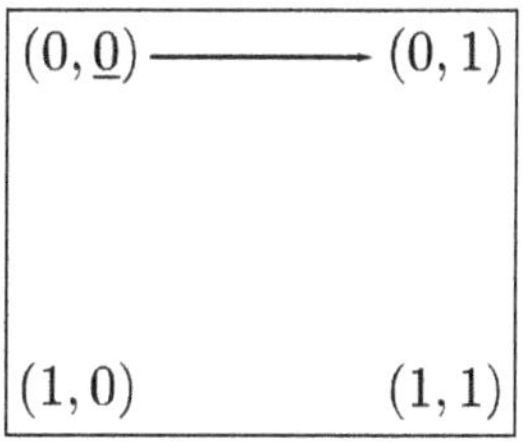

Figure 5.5 The function $\Psi(\mu) = (\mu_1, \overline{\mu_1} \cup \mu_2)$.

and on the other hand

$$(\Psi^{\nu} \circ h)(\mu) = \Psi^{\nu}(\mu_1\overline{\mu_2}, \mu_2) = (\overline{\nu_1}\mu_1\overline{\mu_2} \cup \nu_1\mu_1\overline{\mu_2}, \overline{\nu_2}\mu_2 \cup \nu_2(\overline{\mu_1\overline{\mu_2} \cup \mu_2}))$$
$$= (\mu_1\overline{\mu_2}, \overline{\nu_2}\mu_2 \cup \nu_2(\overline{\mu_1} \cup \mu_2 \cup \mu_2)) = (\mu_1\overline{\mu_2}, \mu_2 \cup \nu_2\overline{\mu_1}).$$

We have proved that $(h, h') : \Phi \longrightarrow \Psi$ is a morphism.

We take in Figure 5.4 $A = \{(0,0), (0,1), (1,1)\}$ and we see that $h(A) = \{(0,0), (0,1)\}$; as h' is bijective, Eq. (5.9) is true and this is remarked in Figure 5.5.

By taking $A = \{(1,0)\}$ in Figure 5.4, we notice that $h(A) = A$. Statement (5.10) is true and statement (5.11) is also true, even if h is not bijective.

5.5 Special Case of Homomorphic Functions vs Invariant Sets

Corollary 5.1 For $\Phi : \mathbf{B}^n \longrightarrow \mathbf{B}^n, A \subset \mathbf{B}^n, A \neq \emptyset$ and $(h, h') \in Hom(\Phi, \Phi)$, if

$$\forall \nu \in \mathbf{B}^n, \Phi^{\nu}(A) \subset A,$$

then

$$\forall \nu \in \mathbf{B}^n, \Phi^{h'(\nu)}(h(A)) \subset h(A);$$

if in addition h' is bijective, then the hypothesis implies the invariance

$$\forall \nu \in \mathbf{B}^n, \Phi^{\nu}(h(A)) \subset h(A).$$

Proof: This follows from Theorem 5.7 for $\Psi = \Phi$ □

Corollary 5.2 If

$$\forall \nu \in \mathbf{B}^n, \Phi^{\nu}(A) = A$$

is true and $(h, h') \in Aut(\Phi)$, then

$$\forall \nu \in \mathbf{B}^n, \Phi^{\nu}(h(A)) = h(A).$$

Proof: This follows from Theorem 5.8 for $\Psi = \Phi$ □

Example 5.6 In Figure 5.3, page 57, the function $\Phi : \mathbf{B}^2 \longrightarrow \mathbf{B}^2$ is defined by $\forall \mu \in \mathbf{B}^2$, $\Phi(\mu_1, \mu_2) = (\mu_1, \overline{\mu_2})$. We take $h : \mathbf{B}^2 \longrightarrow \mathbf{B}^2$ defined by $\forall \mu \in \mathbf{B}^2$,$h(\mu_1, \mu_2) = (\overline{\mu_1}, \mu_2)$. As for arbitrary $\mu, \nu \in \mathbf{B}^2$ we get

$$\Phi^{\nu}(\mu_1, \mu_2) = (\overline{\nu_1}\mu_1 \cup \nu_1\mu_1, \overline{\nu_2}\mu_2 \cup \nu_2\overline{\mu_2}) = (\mu_1, \overline{\nu_2}\mu_2 \cup \nu_2\overline{\mu_2})$$

we can show the fact that $(h, 1_{\mathbf{B}^2}) \in Aut(\Phi)$ in the following way:

$$\begin{aligned} h(\Phi^\nu(\mu)) &= h(\mu_1, \overline{\nu_2}\mu_2 \cup \nu_2\overline{\mu_2}) = (\overline{\mu_1}, \overline{\nu_2}\mu_2 \cup \nu_2\overline{\mu_2}) \\ &= (\overline{\nu_1}\ \overline{\mu_1} \cup \nu_1\overline{\mu_1}, \overline{\nu_2}\mu_2 \cup \nu_2\overline{\mu_2}) \\ &= (\overline{\nu_1}\ \overline{\mu_1} \cup \nu_1\Phi_1(\overline{\mu_1}, \mu_2), \overline{\nu_2}\mu_2 \cup \nu_2\Phi_2(\overline{\mu_1}, \mu_2)) \\ &= \Phi^\nu(\overline{\mu_1}, \mu_2) = \Phi^\nu(h(\mu)). \end{aligned}$$

For $A = \{(0,0),(0,1)\}$ we have $h(A) = \{(1,0),(1,1)\}$ and the invariance of A from the hypothesis of Corollary 5.2 implies the invariance of $h(A)$ from the conclusion of the same corollary. Note that the reasoning is also true if we take $A' = \{(1,0),(1,1)\}$ and $h(A') = \{(0,0),(0,1)\}$.

5.6 Symmetry Relative to Translations vs Invariant Sets

Corollary 5.3 Let $\tau \in \mathbf{B}^n$, $h' : \mathbf{B}^n \longrightarrow \mathbf{B}^n$ such that $(\theta^\tau, h') \in Aut(\Phi)$, $(\theta^\tau, h') \neq (1_{\mathbf{B}^n}, 1_{\mathbf{B}^n})$ and $A \subset \mathbf{B}^n, A \neq \emptyset$.

(a) If

$$\forall \nu \in \mathbf{B}^n, \Phi^\nu(A) \subset A \tag{5.12}$$

then

$$\forall \nu \in \mathbf{B}^n, \Phi^\nu(A \oplus \tau) \subset A \oplus \tau \tag{5.13}$$

where we have denoted

$$A \oplus \tau = \theta^\tau(A);$$

(b) if

$$\forall \nu \in \mathbf{B}^n, \Phi^\nu(A) = A \tag{5.14}$$

then

$$\forall \nu \in \mathbf{B}^n, \Phi^\nu(A \oplus \tau) = A \oplus \tau. \tag{5.15}$$

Proof: Item (a) is a special case of Corollary 5.1, and item (b) is a special case of Corollary 5.2 □

Example 5.7 We can think in Example 5.6 that we have $h(\mu_1, \mu_2) = (\overline{\mu_1}, \mu_2) = \theta^{(1,0)}(\mu_1, \mu_2)$; then $(\theta^{(1,0)}, 1_{\mathbf{B}^2}) \in Aut(\Phi)$.

Example 5.8 We show that the function $\Phi : \mathbf{B}^2 \longrightarrow \mathbf{B}^2, \forall \mu \in \mathbf{B}^2, \Phi(\mu_1, \mu_2) = (\mu_1, \overline{\mu_2})$ from Figure 5.3, page 57 is also symmetrical relative to the translation

$$\forall \mu \in \mathbf{B}^2, \theta^{(1,1)}(\mu) = \mu \oplus (1,1) = (\overline{\mu_1}, \overline{\mu_2}),$$

in the sense that $(\theta^{(1,1)}, 1_{\mathbf{B}^2}) \in Aut(\Phi), (\theta^{(1,1)}, 1_{\mathbf{B}^2}) \neq (1_{\mathbf{B}^2}, 1_{\mathbf{B}^2})$. Indeed, for any $\mu \in \mathbf{B}^2$ and any $\nu \in \mathbf{B}^2$, we can write that

$$\begin{aligned}
(\theta^{(1,1)} \circ \Phi^{\nu})(\mu) &= \theta^{(1,1)}(\overline{\nu_1}\mu_1 \cup \nu_1\mu_1, \overline{\nu_2}\mu_2 \cup \nu_2\overline{\mu_2}) = \theta^{(1,1)}(\mu_1, \overline{\nu_2}\mu_2 \cup \nu_2\overline{\mu_2}) \\
&= (\overline{\mu_1}, \overline{\overline{\nu_2}\mu_2 \cup \nu_2\overline{\mu_2}}) = (\overline{\mu_1}, (\nu_2 \cup \overline{\mu_2})(\overline{\nu_2} \cup \mu_2)) \\
&= (\overline{\mu_1}, \nu_2\mu_2 \cup \overline{\nu_2}\,\overline{\mu_2}) \\
&= (\overline{\nu_1}\,\overline{\mu_1} \cup \nu_1\overline{\mu_1}, \overline{\nu_2}\,\overline{\mu_2} \cup \nu_2\mu_2) = \Phi^{\nu}(\overline{\mu_1}, \overline{\mu_2}) \\
&= (\Phi^{\nu} \circ \theta^{(1,1)})(\mu).
\end{aligned}$$

The hypothesis of the corollary expressed by Eq. (5.14) is fulfilled by the set $A = \{(0,0),(0,1)\}$, thus for $A \oplus (1,1) = \{(1,1),(1,0)\}$ the conclusion (5.15) is fulfilled. We note also the fact that for $A' = \{(1,1),(1,0)\}$ Eq. (5.14) is fulfilled, and this implies the fulfillment of (5.15) by the set $A' \oplus (1,1) = \{(0,0),(0,1)\}$.

5.7 Antihomomorphic Functions vs Invariant Sets

Theorem 5.9 The functions $\Phi, \Psi : \mathbf{B}^n \longrightarrow \mathbf{B}^n$ are given together with $(h,h')^{\frown} \in Hom^{\frown}(\Phi,\Psi)$ and let $A \subset \mathbf{B}^n, A \neq \emptyset$. We suppose that

$$\forall \nu \in \mathbf{B}^n, \Phi^{\nu}(A) = A. \tag{5.16}$$

(a) We have

$$\forall \nu \in \mathbf{B}^n, \Psi^{h'(\nu)}(h(A)) \subset h(A); \tag{5.17}$$

(b) if h' is bijection, then

$$\forall \nu \in \mathbf{B}^n, \Psi^{\nu}(h(A)) \subset h(A); \tag{5.18}$$

(c) if $(h,h')^{\frown} \in Iso^{\frown}(\Phi,\Psi)$, then

$$\forall \nu \in \mathbf{B}^n, \Psi^{\nu}(h(A)) = h(A) \tag{5.19}$$

holds.

Proof: (a) From the hypothesis $(h,h')^{\frown} \in Hom^{\frown}(\Phi,\Psi)$, thus $\forall \nu \in \mathbf{B}^n$ the diagram

$$\begin{array}{ccc}
\mathbf{B}^n & \stackrel{\Phi^{\nu}}{\longrightarrow} & \mathbf{B}^n \\
h \downarrow & & \downarrow h \\
\mathbf{B}^n & \stackrel{\Psi^{h'(\nu)}}{\longleftarrow} & \mathbf{B}^n
\end{array}$$

is commutative. Let $\nu \in \mathbf{B}^n$ and $\mu \in h(A)$ arbitrary. We get the existence of $\lambda \in A$ such that $h(\lambda) = \mu$ and, from the hypothesis (5.16), $\delta \in A$ exists also with $\Phi^{\nu}(\delta) = \lambda$. We infer

$$\Psi^{h'(\nu)}(\mu) = \Psi^{h'(\nu)}(h(\lambda)) = \Psi^{h'(\nu)}(h(\Phi^{\nu}(\delta))) = h(\delta) \in h(A).$$

(b) Obvious.

(c) We take $\nu \in \mathbf{B}^n$, $\mu, \mu' \in h(A)$ arbitrary such that $\mu \neq \mu'$. We have the existence of $\lambda, \lambda' \in A$ with $h(\lambda) = \mu, h(\lambda') = \mu'$ and we have $\lambda \neq \lambda'$, since h is bijective. But (5.16) gives the existence of $\delta, \delta' \in A$ with $\Phi^\nu(\delta) = \lambda, \Phi^\nu(\delta') = \lambda'$ and $\delta \neq \delta'$, because Φ^ν is bijective itself. Therefore

$$\Psi^{h'(\nu)}(\mu) = \Psi^{h'(\nu)}(h(\lambda)) = \Psi^{h'(\nu)}(h(\Phi^\nu(\delta))) = h(\delta)$$
$$\neq h(\delta') = \Psi^{h'(\nu)}(h(\Phi^\nu(\delta'))) = \Psi^{h'(\nu)}(h(\lambda')) = \Psi^{h'(\nu)}(\mu').$$

We conclude that the function $h(A) \ni \mu \mapsto \Psi^{h'(\nu)}(\mu) \in h(A)$ from (b) is injective, thus (5.19) is true □

Remark 5.5 The hypothesis $\forall \nu \in \mathbf{B}^n, \Phi^\nu(A) \subset A$ instead of (5.16) cannot produce a result of the type (5.17) for the following reason. Let $\nu \in \mathbf{B}^n$ arbitrary for which $\Phi^\nu(A) \neq A$ and $\mu \in A \setminus \Phi^\nu(A)$. Then $h(\mu) \in h(A)$ and the situations $\Psi^{h'(\nu)}(h(\mu)) \in h(A), \Psi^{h'(\nu)}(h(\mu)) \notin h(A)$ are both possible.

5.8 Special Case of Antihomomorphic Functions vs Invariant Sets

Corollary 5.4 Given $\Phi : \mathbf{B}^n \longrightarrow \mathbf{B}^n$ and the set $A \subset \mathbf{B}^n, A \neq \emptyset$, we suppose that

$$\forall \nu \in \mathbf{B}^n, \Phi^\nu(A) = A$$

and we take $(h, h')^\frown \in Hom^\frown(\Phi, \Phi)$.

(a) We have

$$\forall \nu \in \mathbf{B}^n, \Phi^{h'(\nu)}(h(A)) \subset h(A);$$

(b) if h' is a bijection, then

$$\forall \nu \in \mathbf{B}^n, \Phi^\nu(h(A)) \subset h(A);$$

(c) if $(h, h')^\frown \in Aut^\frown(\Phi)$, we get

$$\forall \nu \in \mathbf{B}^n, \Phi^\nu(h(A)) = h(A).$$

Proof: This is a consequence of Theorem 5.9 □

Example 5.9 The function from Figure 4.2, page 47 that we have denoted by $\Psi : \forall \mu \in \mathbf{B}^2, \Psi(\mu) = (\overline{\mu_1}, \mu_2)$ fulfills $(h, 1_{\mathbf{B}^2})^\frown \in Aut^\frown(\Psi)$, where $h : \mathbf{B}^2 \longrightarrow \mathbf{B}^2$ is given by $\forall \mu \in \mathbf{B}^2, h(\mu) = (\mu_1, \overline{\mu_2})$. In order to prove this assertion, we take some $\mu, \nu \in \mathbf{B}^2$ resulting:

$$\Psi^\nu(\mu) = (\overline{\nu_1}\mu_1 \cup \nu_1\overline{\mu_1}, \overline{\nu_2}\mu_2 \cup \nu_2\mu_2) = (\overline{\nu_1}\mu_1 \cup \nu_1\overline{\mu_1}, \mu_2),$$
$$\Psi^\nu(h(\Psi^\nu(\mu))) = \Psi^\nu(h(\overline{\nu_1}\mu_1 \cup \nu_1\overline{\mu_1}, \mu_2)) = \Psi^\nu(\overline{\nu_1}\mu_1 \cup \nu_1\overline{\mu_1}, \overline{\mu_2})$$

$$
\begin{aligned}
&= (\overline{\nu_1}(\overline{\nu_1}\mu_1 \cup \nu_1\overline{\mu_1}) \cup \nu_1\Psi_1(\overline{\nu_1}\mu_1 \cup \nu_1\overline{\mu_1}, \overline{\mu_2}),\\
&\quad \overline{\nu_2}\ \overline{\mu_2} \cup \nu_2\Psi_2(\overline{\nu_1}\mu_1 \cup \nu_1\overline{\mu_1}, \overline{\mu_2}))\\
&= (\overline{\nu_1}\mu_1 \cup \nu_1\overline{\overline{\nu_1}\mu_1 \cup \nu_1\overline{\mu_1}}, \overline{\nu_2}\ \overline{\mu_2} \cup \nu_2\overline{\mu_2})\\
&= (\overline{\nu_1}\mu_1 \cup \nu_1(\nu_1 \cup \overline{\mu_1})(\overline{\nu_1} \cup \mu_1), \overline{\mu_2})\\
&= (\overline{\nu_1}\mu_1 \cup \nu_1(\nu_1 \cup \overline{\mu_1})\mu_1, \overline{\mu_2}) = (\overline{\nu_1}\mu_1 \cup \nu_1\mu_1, \overline{\mu_2})\\
&= (\mu_1, \overline{\mu_2}) = h(\mu).
\end{aligned}
$$

We denote $A = \{(0,0),(1,0)\}$, thus $h(A) = \{(0,1),(1,1)\}$. We have $\forall \nu \in \mathbf{B}^2$, $\Psi^\nu(A) = A$ indeed, wherefrom $\forall \nu \in \mathbf{B}^2, \Psi^\nu(h(A)) = h(A)$.

5.9 Antisymmetry Relative to Translations vs Invariant Sets

Corollary 5.5 The function $\Phi : \mathbf{B}^n \longrightarrow \mathbf{B}^n$ and $A \subset \mathbf{B}^n, A \neq \emptyset$ are given. We suppose that

$$\forall \nu \in \mathbf{B}^n, \Phi^\nu(A) = A \tag{5.20}$$

and $\tau \in \mathbf{B}^n$, $h' : \mathbf{B}^n \longrightarrow \mathbf{B}^n$ exist such that $(\theta^\tau, h')^{\frown} \in Aut^{\frown}(\Phi)$. Then

$$\forall \nu \in \mathbf{B}^n, \Phi^\nu(A \oplus \tau) = A \oplus \tau \tag{5.21}$$

where we have used the notation

$$A \oplus \tau = \{\mu \oplus \tau | \mu \in A\}.$$

Proof: This is a special case of Corollary 5.4 (c), when h is the translation with τ □

Example 5.10 We get back to the function $\Phi : \mathbf{B}^2 \longrightarrow \mathbf{B}^2$, $\forall \mu \in \mathbf{B}^2$, $\Phi(\mu_1, \mu_2) = (\mu_1, \overline{\mu_2})$ from Figure 5.3, page 57 and we show that $(\theta^{(1,1)}, 1_{\mathbf{B}^2})^{\frown} \in Aut^{\frown}(\Phi)$: for any $\mu \in \mathbf{B}^2$ and any $\nu \in \mathbf{B}^2$ we infer

$$
\begin{aligned}
(\Phi^\nu \circ \theta^{(1,1)} \circ \Phi^\nu)(\mu) &= (\Phi^\nu \circ \theta^{(1,1)})(\mu_1, \overline{\nu_2}\mu_2 \cup \nu_2\overline{\mu_2})\\
&= \Phi^\nu(\overline{\mu_1}, \overline{\overline{\nu_2}\mu_2 \cup \nu_2\overline{\mu_2}})\\
&= \Phi^\nu(\overline{\mu_1}, (\nu_2 \cup \overline{\mu_2})(\overline{\nu_2} \cup \mu_2)) = \Phi^\nu(\overline{\mu_1}, \nu_2\mu_2 \cup \overline{\nu_2}\ \overline{\mu_2})\\
&= (\overline{\mu_1}, \overline{\nu_2}(\nu_2\mu_2 \cup \overline{\nu_2}\ \overline{\mu_2}) \cup \nu_2\overline{\nu_2\mu_2 \cup \overline{\nu_2}\ \overline{\mu_2}})\\
&= (\overline{\mu_1}, \overline{\nu_2}\ \overline{\mu_2} \cup \nu_2(\overline{\nu_2} \cup \overline{\mu_2})(\nu_2 \cup \mu_2))\\
&= (\overline{\mu_1}, \overline{\nu_2}\ \overline{\mu_2} \cup \nu_2(\overline{\nu_2}\mu_2 \cup \nu_2\overline{\mu_2}))\\
&= (\overline{\mu_1}, \overline{\nu_2}\ \overline{\mu_2} \cup \nu_2\overline{\mu_2}) = (\overline{\mu_1}, \overline{\mu_2}) = \theta^{(1,1)}(\mu).
\end{aligned}
$$

Similarly with the analysis that was made at Example 5.8, page 63 the hypothesis of the corollary expressed by Eq. (5.20) is satisfied by the set $A = \{(0,0),(0,1)\}$, therefore the conclusion (5.21) is satisfied by $A \oplus (1,1) = \{(1,1),(1,0)\}$; in addition, for $A' = \{(1,1),(1,0)\}$ Eq. (5.20) is true and this implies the satisfaction of (5.21) by the set $A' \oplus (1,1) = \{(0,0),(0,1)\}$.

5.10 Relatively Isolated Sets, Isolated Set

Definition 5.2 We consider the function $\Phi : \mathbf{B}^n \longrightarrow \mathbf{B}^n$ and the nonempty sets $A_1, \dots, A_p \subset \mathbf{B}^n, p \in \{2, \dots, 2^n\}$ that are disjoint two by two:

$$\forall i \in \{1, \dots, p\}, \forall j \in \{1, \dots, p\}, i \neq j \Longrightarrow A_i \cap A_j = \emptyset. \tag{5.22}$$

If one of

$$\forall i \in \{1, \dots, p\}, \forall \nu \in \mathbf{B}^n, \Phi^\nu(A_i) \subset A_i, \tag{5.23}$$

$$\forall i \in \{1, \dots, p\}, \forall \nu \in \mathbf{B}^n, \Phi^\nu(A_i) = A_i \tag{5.24}$$

is true, then $A_1, \dots, A_p$ are called **k-relatively isolated,** $k \in \{(5.23), (5.24)\}$. In particular, if $\emptyset \subsetneq A \subsetneq \mathbf{B}^n$ and A, $\mathbf{B}^n \backslash A$ are k-relatively isolated, we say that A is **k-isolated.**

Example 5.11 In Figure 5.1, page 57 there are no k-isolated sets, $k \in \{(5.23), (5.24)\}$.

Example 5.12 In Figure 5.2, page 57 the sets $\{(0,0)\}$ and $\{(1,0),(1,1),(0,1)\}$ are (5.23)-isolated.

Example 5.13 In Figure 5.3, page 57 the sets $\{(0,0),(0,1)\}$ and $\{(1,0),(1,1)\}$ are k-isolated, $k \in \{(5.23), (5.24)\}$.

Theorem 5.10 If $\mu \in \mathbf{B}^n$ is an isolated fixed point of $\Phi : \mathbf{B}^n \longrightarrow \mathbf{B}^n$, then $\{\mu\}$ is k-isolated, $k \in \{(5.23), (5.24)\}$.

Proof: We choose to make the proof for (5.23). The hypothesis states that $\mu^- = \mu^+ = \{\mu\}$. From $\mu^+ = \{\mu\}$ we get $\forall \nu \in \mathbf{B}^n, \Phi^\nu(\mu) = \mu$, thus the invariance $\Phi^\nu(\{\mu\}) \subset \{\mu\}$ holds. From $\mu^- = \{\mu\}$, we have

$$\forall \nu \in \mathbf{B}^n, \Phi^\nu(\mathbf{B}^n \backslash \{\mu\}) \subset \mathbf{B}^n \backslash \{\mu\}; \tag{5.25}$$

indeed, (5.25) is equivalent with any of

$$\forall \nu \in \mathbf{B}^n, \forall \mu' \in \mathbf{B}^n, \mu \neq \mu' \Longrightarrow \Phi^\nu(\mu') \neq \mu,$$

$$\forall \nu \in \mathbf{B}^n, \forall \mu' \in \mathbf{B}^n, \mu = \mu' \text{ or } \Phi^\nu(\mu') \neq \mu$$

and we suppose against all reason the falsity of the last statement. We have then the existence of $\nu \in \mathbf{B}^n$ and $\mu' \in \mathbf{B}^n$, $\mu' \neq \mu$ with $\Phi^\nu(\mu') = \mu$, i.e. $\mu' \in \mu^-$, contradiction. Statement (5.25) is true, thus $\{\mu\}$ is (5.23)-isolated □

Theorem 5.11 If the sets $A_1, \dots, A_p \subset \mathbf{B}^n, p \in \{3, \dots, 2^n\}$ are k-relatively isolated, $k \in \{(5.23), (5.24)\}$, then $A_1 \cup A_2, A_3, \dots, A_p$ are k-relatively isolated. In particular, if $A, B, \mathbf{B}^n \setminus (A \cup B)$ are k-relatively isolated, then $A \cup B$ is k-isolated.

Proof: We choose to refer to $k = (5.23)$. In order to see the first statement, we notice that $A_1 \cup A_2, A_3, \dots, A_p$ are disjoint two by two. The fact that

$$\forall \nu \in \mathbf{B}^n, \Phi^\nu(A_1 \cup A_2) \subset A_1 \cup A_2$$

has been proved at Theorem 5.6, page 59 □

5.11 Isomorphic Functions vs Relatively Isolated Sets

Theorem 5.12 Let the functions $\Phi, \Psi : \mathbf{B}^n \longrightarrow \mathbf{B}^n$, the (5.23)-relatively isolated sets $A_1, \dots, A_p \subset \mathbf{B}^n, p \in \{2, \dots, 2^n\}$, in the sense that (5.22), (5.23) are true, and the isomorphism $(h, h') \in Iso(\Phi, \Psi)$. Then the sets $h(A_1), \dots, h(A_k)$ are (5.23)-relatively isolated, i.e.

$$\forall i \in \{1, \dots, p\}, \forall j \in \{1, \dots, p\}, i \neq j \Longrightarrow h(A_i) \cap h(A_j) = \emptyset, \tag{5.26}$$

$$\forall i \in \{1, \dots, p\}, \forall \nu \in \mathbf{B}^n, \Psi^{\nu}(h(A_i)) \subset h(A_i) \tag{5.27}$$

hold. Moreover, if (5.23) is replaced by (5.24), then (5.27) is replaced by

$$\forall i \in \{1, \dots, p\}, \forall \nu \in \mathbf{B}^n, \Psi^{\nu}(h(A_i)) = h(A_i). \tag{5.28}$$

Proof: We suppose against all reason that (5.26) is false, i.e.

$$\exists i \in \{1, \dots, p\}, \exists j \in \{1, \dots, p\}, i \neq j \text{ and } h(A_i) \cap h(A_j) \neq \emptyset$$

and let $\mu \in h(A_i) \cap h(A_j)$. Then $\mu^i \in A_i$ and $\mu^j \in A_j$ exist such that

$$h(\mu^i) = \mu = h(\mu^j). \tag{5.29}$$

But $A_i \cap A_j = \emptyset$ implies $\mu^i \neq \mu^j$. We have obtained a contradiction between (5.29) and the bijectivity of h. (5.26) is true.

Statement (5.27) is a consequence of Theorem 5.7, page 60 and statement (5.28) results from Theorem 5.8, page 60 □

5.12 Antiisomorphic Functions vs Relatively Isolated Sets

Theorem 5.13 Let the functions $\Phi, \Psi : \mathbf{B}^n \longrightarrow \mathbf{B}^n$, the (5.24)-relatively isolated sets $A_1, \dots, A_p \subset \mathbf{B}^n, p \in \{2, \dots, 2^n\}$ and the antiisomorphism $(h, h')^{\frown} \in Iso^{\frown}(\Phi, \Psi)$. Then $h(A_1), \dots, h(A_p)$ are (5.24)-relatively isolated, in the sense that

$$\forall i \in \{1, \dots, p\}, \forall j \in \{1, \dots, p\}, i \neq j \Longrightarrow h(A_i) \cap h(A_j) = \emptyset, \tag{5.30}$$

$$\forall i \in \{1, \dots, p\}, \forall v \in \mathbf{B}^n, \Psi^v(h(A_i)) = h(A_i) \tag{5.31}$$

hold.

Proof: The truth of (5.30) is proved similarly with proof of (5.26) from Theorem 5.12. Statement (5.31) follows from Theorem 5.9, page 64 □

6

Invariant Subsets

Let $\emptyset \neq A \subset X \subset \mathbf{B}^n$ and $\Phi : \mathbf{B}^n \longrightarrow \mathbf{B}^n$. If A is invariant, then it is called an invariant subset of X.

The morphisms and the antimorphisms bring invariant subsets in invariant subsets.

If X has invariant subsets, then their union is the maximal invariant subset of X. At the same time, if X has invariant subsets and if their intersection A is nonempty, then A is the minimal invariant subset of X.

An invariant set X is disconnected if it has a proper invariant subset, and it is connected (or minimal) otherwise. The connected components $X_1, \dots, X_p$ of X are a partition of connected sets.

6.1 Definition

Definition 6.1   We consider $\emptyset \neq A \subset X \subset \mathbf{B}^n$ and $\Phi : \mathbf{B}^n \longrightarrow \mathbf{B}^n$. If one of

$$\forall v \in \mathbf{B}^n, \Phi^v(A) \subset A, \tag{6.1}$$

$$\forall v \in \mathbf{B}^n, \Phi^v(A) = A \tag{6.2}$$

is true, then A is called a k-**invariant subset** of X, $k \in \{(6.1), (6.2)\}$.

Remark 6.1   The concept of invariance of a subset allows studying, for example the situation when X is not invariant and A is the greatest k-invariant subset of X, $k \in \{(6.1), (6.2)\}$. We can also think of situations like

$$\begin{cases} \forall v \in \mathbf{B}^n, \Phi^v(A) = A, \\ \forall v \in \mathbf{B}^n, \Phi^v(X) \subset X \end{cases} \tag{6.3}$$

when the properties of invariance of A and X differ.

Boolean Functions: Topics in Asynchronicity, First Edition. Serban E. Vlad.
© 2019 John Wiley & Sons, Inc. Published 2019 by John Wiley & Sons, Inc.

Remark 6.2 Note that the possibility when the invariance of A is weaker than the invariance of X:

$$\begin{cases} \forall v \in \mathbf{B}^n, \Phi^v(A) \subset A, \\ \exists v \in \mathbf{B}^n, \Phi^v(A) \subsetneq A, \\ \forall v \in \mathbf{B}^n, \Phi^v(X) = X \end{cases}$$

does not exist. If the previous three statements would be true, then $v \in \mathbf{B}^n$ would exist such that the restriction of Φ^v at A would be injective and non-injective at the same time.

6.2 Examples

Example 6.1 If X is k-invariant, for $k \in \{(6.1), (6.2)\}$, then it is also a k-invariant subset of itself.

Example 6.2 Any k-invariant set $A \subset \mathbf{B}^n$, $k \in \{(6.1), (6.2)\}$ is a k-invariant subset of $X = \mathbf{B}^n$.

Example 6.3 In Figure 5.4, page 61, the subsets $\{(0,0), (0,1), (1,1)\}$, $\{(0,1), (1,1)\}$ of $X = \{(0,0), (0,1), (1,1)\}$ are (6.1)-invariant, and $\{(1,1)\} \subset X$ is (6.2)-invariant.

Example 6.4 If $\mu \in X$ fulfills $\Phi(\mu) = \mu$, then $A = \{\mu\}$ is a (6.2)-invariant subset of X. Moreover, a subset $A \subset X$ of fixed points of Φ is a (6.2)-invariant subset of X.

Example 6.5 In Figure 5.3, page 57, the sets $A = \mathbf{B}^2, A' = \{(0,0), (0,1)\}, A'' = \{(1,0), (1,1)\}$ are (6.2)-invariant subsets of $X = \mathbf{B}^2$.

6.3 Maximal Invariant Subset

Definition 6.2 Let $X \subset \mathbf{B}^n$ arbitrary and $A \subset X, A \neq \emptyset$. A is called the **maximal invariant subset** of X if one of

$$\begin{cases} \forall v \in \mathbf{B}^n, \Phi^v(A) \subset A, \\ \forall Y, (\emptyset \subsetneq Y \subset X \text{ and } \forall v \in \mathbf{B}^n, \Phi^v(Y) \subset Y) \Longrightarrow Y \subset A, \end{cases} \tag{6.4}$$

$$\begin{cases} \forall v \in \mathbf{B}^n, \Phi^v(A) = A, \\ \forall Y, (\emptyset \subsetneq Y \subset X \text{ and } \forall v \in \mathbf{B}^n, \Phi^v(Y) = Y) \Longrightarrow Y \subset A \end{cases} \tag{6.5}$$

holds. In these situations, we refer to the maximal k-invariant subset of X, $k \in \{(6.1),(6.2)\}$.

Example 6.6 We take in Figure 5.5, page 61 $X = \{(0,0),(0,1),(1,1)\}$. Then X is (6.1)-invariant, i.e. $\forall v \in \mathbf{B}^n, \Phi^v(X) \subset X$, thus it is the maximal (6.1)-invariant subset of X. On the other hand, the (6.2)-invariant subsets of X are $A = \{(0,1)\}$, $A' = \{(1,1)\}$, $A'' = \{(0,1),(1,1)\}$ and A'' is the maximal (6.2)-invariant subset of X.

Example 6.7 We take in Figure 3.2, page 37 $X = \{(0,0),(0,1)\}$. Then X does not fulfill any property of invariance and the only invariant subset of X is $A = \{(0,1)\}$ ((6.2) is true), which is also maximal.

Theorem 6.1 For $X \subset \mathbf{B}^n, X \neq \emptyset$ and $k \in \{(6.1),(6.2)\}$, the following properties hold:

(a) if X has a k-invariant subset, then it has a maximal k-invariant subset;

(b) if X is k-invariant, then it is the maximal k-invariant subset of itself.

Proof: In order to make a choice, we refer to $k = (6.1)$.

(a) Let $A_1, \ldots, A_p \subset X$ be the nonempty (6.1)-invariant subsets of X. In this situation, Theorem 5.6, page 59 states that the nonempty set $A_1 \cup \ldots \cup A_p \subset X$ is (6.1)-invariant. We have $A_i \subset A_1 \cup \ldots \cup A_p, i \in \{1, \ldots, p\}$.

(b) The set X is the maximal subset of itself relative to the inclusion; if X is (6.1)-invariant, then it is the maximal (6.1)-invariant subset of itself □

Theorem 6.2 We consider the functions $\Phi, \Psi : \mathbf{B}^n \longrightarrow \mathbf{B}^n$, the sets $\emptyset \subsetneq A \subset X \subset \mathbf{B}^n$ and $(h, h') \in Iso(\Phi, \Psi)$. We have $\emptyset \subsetneq h(A) \subset h(X) \subset \mathbf{B}^n$ and

(a) if the maximality property (6.4) of A holds, then the maximality property

$$\begin{cases} \forall v \in \mathbf{B}^n, \Psi^v(h(A)) \subset h(A), \\ \forall Y, (\emptyset \subsetneq Y \subset h(X) \text{ and } \forall v \in \mathbf{B}^n, \Psi^v(Y) \subset Y) \Longrightarrow Y \subset h(A) \end{cases} \tag{6.6}$$

of $h(A)$ holds also;

(b) if the maximality property (6.5) of A holds, then the maximality property

$$\begin{cases} \forall v \in \mathbf{B}^n, \Psi^v(h(A)) = h(A), \\ \forall Y, (\emptyset \subsetneq Y \subset h(X) \text{ and } \forall v \in \mathbf{B}^n, \Psi^v(Y) = Y) \Longrightarrow Y \subset h(A) \end{cases} \tag{6.7}$$

of $h(A)$ holds too.

Proof: (a) By Theorem 5.7, page 60, the (6.1)-invariance of A relative to Φ implies the (6.1)-invariance of $h(A)$ relative to Ψ.

Let now Y arbitrary such that $\emptyset \subsetneq Y \subset h(X)$ and Y is (6.1)-invariant relative to Ψ. As $(h^{-1}, h'^{-1}) \in Iso(\Psi, \Phi)$, we get from Theorem 5.7 that $h^{-1}(Y)$ satisfies

$\emptyset \subsetneq h^{-1}(Y) \subset X$, and moreover it is (6.1)-invariant relative to Φ. The hypothesis (6.4) implies that $h^{-1}(Y) \subset A$, in other words $Y \subset h(A)$, therefore $h(A)$ is the maximal (6.1)-invariant subset of $h(X)$ relative to Ψ.

(b) The proof is similar with the proof of (a) by replacing Theorem 5.7 with Theorem 5.8, page 60 □

Theorem 6.3 The functions $\Phi, \Psi : \mathbf{B}^n \longrightarrow \mathbf{B}^n$ and the sets $\emptyset \subsetneq A \subset X \subset \mathbf{B}^n$ are given, together with the antiisomorphism $(h, h')^{\smile} \in Iso^{\smile}(\Phi, \Psi)$. We have $\emptyset \subsetneq h(A) \subset h(X) \subset \mathbf{B}^n$ and if (6.5) is true, then the maximality (6.7) of $h(A)$ is true.

Proof: The fact that A is (6.2)-invariant relative to Φ makes from Theorem 5.9, page 64 that $h(A)$ is (6.2)-invariant relative to Ψ.

Let Y a nonempty arbitrary set such that $Y \subset h(X)$ and Y is (6.2)-invariant relative to Ψ. Then $h^{-1}(Y) \subset X$ is nonempty and (6.2)-invariant relative to Φ, thus in accordance with the hypothesis (6.5), $h^{-1}(Y) \subset A$. We infer from here that $Y \subset h(A)$, i.e. $h(A)$ is the maximal (6.2)-invariant subset of $h(X)$ relative to Ψ □

6.4 Minimal Invariant Subset

Definition 6.3 Let $X \subset \mathbf{B}^n$ and $A \subset X$ nonempty. If one of

$$\begin{cases} \forall \nu \in \mathbf{B}^n, \Phi^{\nu}(A) \subset A, \\ \forall Y, (\emptyset \subsetneq Y \subset X \text{ and } \forall \nu \in \mathbf{B}^n, \Phi^{\nu}(Y) \subset Y) \Longrightarrow A \subset Y, \end{cases} \tag{6.8}$$

$$\begin{cases} \forall \nu \in \mathbf{B}^n, \Phi^{\nu}(A) = A, \\ \forall Y, (\emptyset \subsetneq Y \subset X \text{ and } \forall \nu \in \mathbf{B}^n, \Phi^{\nu}(Y) = Y) \Longrightarrow A \subset Y \end{cases} \tag{6.9}$$

holds, we use to say that A is the **minimal k-invariant subset** of X, $k \in \{(6.1), (6.2)\}$.

Definition 6.4 If X is the minimal invariant subset of itself:

$$\begin{cases} \forall \nu \in \mathbf{B}^n, \Phi^{\nu}(X) \subset X, \\ \forall Y, (\emptyset \subsetneq Y \subset X \text{ and } \forall \nu \in \mathbf{B}^n, \Phi^{\nu}(Y) \subset Y) \Longrightarrow X \subset Y, \end{cases} \tag{6.10}$$

$$\begin{cases} \forall \nu \in \mathbf{B}^n, \Phi^{\nu}(X) = X, \\ \forall Y, (\emptyset \subsetneq Y \subset X \text{ and } \forall \nu \in \mathbf{B}^n, \Phi^{\nu}(Y) = Y) \Longrightarrow X \subset Y \end{cases} \tag{6.11}$$

then it is called **k-minimal** or **k-connected,** $k \in \{(6.1), (6.2)\}$.

Remark 6.3 Definition 6.3 states nothing about the invariance of X, while in Definition 6.4 the invariance of X is essential.

Remark 6.4 If X has invariant subsets and if the intersection A of these sets is nonempty, then A is invariant and moreover it is the minimal invariant subset of X.

Remark 6.5 Statements (6.10) and (6.11) may be written under the equivalent form

$$\begin{cases} \forall v \in \mathbf{B}^n, \Phi^v(X) \subset X, \\ \forall Y, (\emptyset \subsetneq Y \subset X \text{ and } \forall v \in \mathbf{B}^n, \Phi^v(Y) \subset Y) \Longrightarrow X = Y, \end{cases} \tag{6.12}$$

$$\begin{cases} \forall v \in \mathbf{B}^n, \Phi^v(X) = X, \\ \forall Y, (\emptyset \subsetneq Y \subset X \text{ and } \forall v \in \mathbf{B}^n, \Phi^v(Y) = Y) \Longrightarrow X = Y. \end{cases} \tag{6.13}$$

In order to notice the equivalence between the second statement (6.12) and the second statement (6.10), let Y arbitrary, fixed. We obtain the equivalent statements in succession:

$$\text{not } (Y \neq \emptyset \text{ and } Y \subset X \text{ and } \forall v \in \mathbf{B}^n, \Phi^v(Y) \subset Y) \text{ or } (X \subset Y \text{ and } Y \subset X),$$
$$(Y = \emptyset \text{ or } Y \not\subset X \text{ or } \exists v \in \mathbf{B}^n, \Phi^v(Y) \not\subset Y) \text{ or } (X \subset Y \text{ and } Y \subset X),$$

etc.

Example 6.8 For the function $1_{\mathbf{B}^n}$ and $X = \mathbf{B}^n$, any $A \subset X, A \neq \emptyset$ is a (6.2)-invariant subset (Example 5.1, page 57) and X has no minimal invariant subset ((6.8), (6.9) are false, both of them, for any $A \subset \mathbf{B}^n$).

Example 6.9 In Figure 5.2, page 57 the set $X = \{(1,0),(1,1),(0,1)\}$ is (6.1)-invariant, the subsets $A = X, A' = \{(1,1),(0,1)\}, A'' = \{(0,1)\}$ are all (6.1)-invariant and A'' is the minimal (6.1)-invariant subset of X (it fulfills (6.8)). As $(0,1)$ is a fixed point, A'' is also (6.2)-invariant; it is the only subset with this property, therefore it is the minimal (6.2)-invariant subset of X.

Example 6.10 In Figure 5.1, page 57 $X = \mathbf{B}^2$ is (6.1)-minimal, i.e. it is (6.1)-invariant and it has no proper (6.1)-invariant subset.

Example 6.11 In Figure 5.3, page 57:

- $X = \{(0,0)\}$ is not k-invariant and it has no k-invariant subset, $k \in \{(6.1),(6.2)\}$.
- $X = \{(0,0),(0,1),(1,0)\}$ is not k-invariant, $k \in \{(6.1),(6.2)\}$, $A = \{(0,0),(0,1)\}$ is an invariant (6.2)-subset and it is the minimal (6.2)-invariant subset;
- $X = \{(0,0),(0,1)\}$ is (6.2)-invariant without proper k-invariant subsets, $k \in \{(6.1),(6.2)\}$, hence X is (6.2)-minimal;
- $X = \mathbf{B}^2$ is (6.2)-invariant and a minimal k-invariant subset does not exist, $k \in \{(6.1),(6.2)\}$ since $A' = \{(0,0),(0,1)\}, A'' = \{(1,0),(1,1)\}$ are both (6.2)-invariant and disjoint.

Remark 6.6 Statements dual to Theorems 6.2 and 6.3 hold for the minimality properties (6.8), (6.9). In particular, isomorphisms $(h, h') : \Phi \longrightarrow \Psi$ and anti-isomorphisms $(h, h')^{\frown} : \Phi \longrightarrow \Psi$ bring the minimal set X of Φ in the minimal set $h(X)$ of Ψ.

6.5 Connected Components

Theorem 6.4 Let $X \subset \mathbf{B}^n$ nonempty and $X_1, \dots, X_p \subset X$, $A_1, \dots, A_q \subset X$ two partitions of k-connected subsets of X, $k \in \{(6.1), (6.2)\}$. Then $p = q$ and we have, modulo the order of these sets, $X_1 = A_1, \dots, X_p = A_p$.

Proof: For $k = (6.1)$, the hypothesis states that $\forall i \in \{1, \dots, p\}$,

$$\begin{cases} \forall v \in \mathbf{B}^n, \Phi^v(X_i) \subset X_i, \\ \forall Y, (\emptyset \subsetneq Y \subset X_i \text{ and } \forall v \in \mathbf{B}^n, \Phi^v(Y) \subset Y) \Longrightarrow X_i = Y, \end{cases} \tag{6.14}$$

and $\forall j \in \{1, \dots, q\}$,

$$\begin{cases} \forall v \in \mathbf{B}^n, \Phi^v(A_j) \subset A_j, \\ \forall Y, (\emptyset \subsetneq Y \subset A_j \text{ and } \forall v \in \mathbf{B}^n, \Phi^v(Y) \subset Y) \Longrightarrow A_j = Y \end{cases} \tag{6.15}$$

hold. The set X_1 overlaps with one of $A_1, \dots, A_q$ and we can suppose without loosing the generality that $X_1 \cap A_1 \neq \emptyset$. We infer that $X_1 \cap A_1$ is a (6.1)-invariant subset of X_1, also a (6.1)-invariant subset of A_1 and

$$X_1 \overset{(6.14)}{=} X_1 \cap A_1 \overset{(6.15)}{=} A_1$$

etc □

Definition 6.5 The sets $X_1, \dots, X_p$ that are k-connected subsets of X, $k \in \{(6.1), (6.2)\}$, are called the k-**connected components** of X.

Remark 6.7 The connected components of X are a special case of relatively isolated sets, see Definition 5.2, page 67, namely $X_1, \dots, X_p$ are asked to be all minimal (without proper invariant subsets).

Example 6.12 In Figure 6.1, we see that the (6.1)-connected components of $\mathbf{B}^3$ are $X_1 = \{(0,0,0), (0,0,1), (0,1,1), (0,1,0)\}$, $X_2 = \{(1,0,0), (1,0,1), (1,1,1), (1,1,0)\}$. In this example (6.2)-connected components do not exist.

Example 6.13 The function from Figure 6.2 has the property that (6.2)-connected components of $\mathbf{B}^2$ exist, $X_1 = \{(0,0), (0,1)\}$ and $X_2 = \{(1,0), (1,1)\}$. They are (6.1)-connected components of $\mathbf{B}^2$ also.

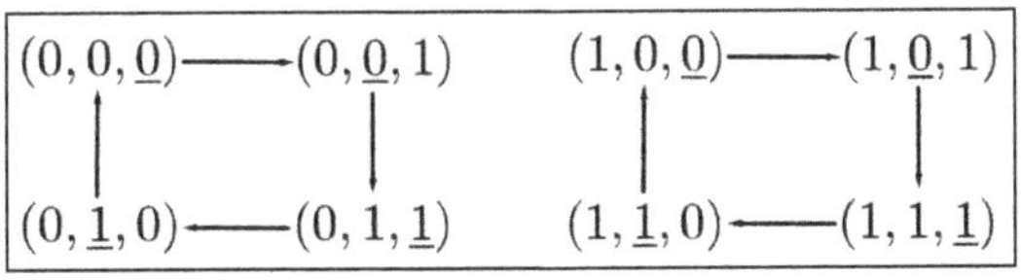

Figure 6.1 The sets $X_1 = \{(0,0,0),(0,0,1),(0,1,1),(0,1,0)\}$ and $X_2 = \{(1,0,0),(1,0,1),(1,1,1),(1,1,0)\}$ are the (6.1)-connected components of $\mathbf{B}^3$.

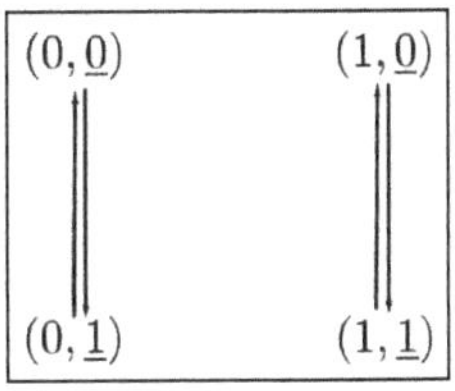

Figure 6.2 The sets $X_1 = \{(0,0),(0,1)\}$ and $X_2 = \{(1,0),(1,1)\}$ are the (6.2)-connected components of $\mathbf{B}^2$.

Remark 6.8 The bijections $h : \mathbf{B}^n \longrightarrow \mathbf{B}^n$ bring a partition $X_1, \dots, X_p$, $p \geq 2$ of X in a partition $h(X_1), \dots, h(X_p)$ of $h(X)$. We get that the isomorphisms $(h,h') : \Phi \longrightarrow \Psi$ and the antiisomorphisms $(h,h')^{\smallfrown} : \Phi \longrightarrow \Psi$ bring k-connected components of X relative to Φ in k-connected components of $h(X)$ relative to Ψ, in a manner suggested by Theorems 6.2 and 6.3, page 73.

6.6 Disconnected Set

Definition 6.6 Let $\Phi : \mathbf{B}^n \longrightarrow \mathbf{B}^n$ and the nonempty subset $X \subset \mathbf{B}^n$. If either of

$$\begin{cases} \forall \nu \in \mathbf{B}^n, \Phi^\nu(X) \subset X, \\ \exists Y, \emptyset \subsetneq Y \subsetneq X \text{ and } \forall \nu \in \mathbf{B}^n, \Phi^\nu(Y) \subset Y, \end{cases} \tag{6.16}$$

$$\begin{cases} \forall \nu \in \mathbf{B}^n, \Phi^\nu(X) = X, \\ \exists Y, \emptyset \subsetneq Y \subsetneq X \text{ and } \forall \nu \in \mathbf{B}^n, \Phi^\nu(Y) = Y \end{cases} \tag{6.17}$$

holds, we say that the set X is k-**disconnected** or k- **separated** and each Y like previously is called a k- **separation** of X, $k \in \{(6.1),(6.2)\}$. We use to say that X is k-**separated** by Y.

Remark 6.9 The negation of the second statement of (6.12) is equivalent with the second statement of (6.16) and the negation of the second statement of (6.13) is equivalent with the second statement of (6.17).

Remark 6.10 We notice that the k-invariant set X is k-connected if it has no proper k-invariant subset, see Definition 6.4, page 74.

The k-invariant set X is k-disconnected, otherwise, i.e. if it has a proper k-invariant subset Y; thus, there is no connection from Y to $X\backslash Y$.

Remark 6.11 We suppose that $Y \subset \mathbf{B}^n$ is $(5.23)_{page\ 67}$ isolated, see Definition 5.2, page 67. Then it is a (6.1)-separation of $\mathbf{B}^n$, thus $\mathbf{B}^n$ is (6.1)-disconnected. In this case, there is no connection from Y to $\mathbf{B}^n\backslash Y$ and also no connection from $\mathbf{B}^n\backslash Y$ to Y. The same is true for Y $(5.24)_{page\ 67}$-isolated and (6.2)-separation of $\mathbf{B}^n$.

Theorem 6.5 We suppose that X, Y, Z are k-invariant, $k \in \{(6.1), (6.2)\}$, where $\emptyset \subsetneq Y \subsetneq X$, i.e. Y is a k-separation of X, and $Z \subset X$. If Z is k-minimal, then

$$Z \subset Y \text{or } Z \subset X\backslash Y \tag{6.18}$$

holds.

Proof: We fix $k \in \{(6.1), (6.2)\}$ arbitrary. If, against all reason, (6.18) is false, then

$$Z \cap Y \neq \emptyset, \tag{6.19}$$

$$Z \cap (X\backslash Y) \neq \emptyset \tag{6.20}$$

are true. In this situation, Theorem 5.6, page 59 and (6.19) show that $Z \cap Y$ is k-invariant. Moreover, (6.19) and (6.20) show that $Z \cap Y \subsetneq Z$. We have obtained a contradiction with the supposition that Z is k-minimal □

Theorem 6.6 If $X, X' \subset \mathbf{B}^n$ nonempty are k-invariant, $k \in \{(6.1), (6.2)\}$ and Y is a k-separation of X, then Y is a k-separation of $X \cup X'$.

Proof: Let $k \in \{(6.1), (6.2)\}$ arbitrary. The hypothesis states that Y is k-invariant and in addition $\emptyset \subsetneq Y \subsetneq X$. We have $\emptyset \subsetneq Y \subsetneq X \cup X'$ and, as $X \cup X'$ is k-invariant, Y is a k-separation of $X \cup X'$ □

Theorem 6.7 (a) If X, X' are k-invariant, $k \in \{(6.1), (6.2)\}$ and $X \subsetneq X \cup X'$, then X is a k-separation of $X \cup X'$.

(b) If in addition X, X' are k-connected, then $X \cup X'$ is k-disconnected.

Proof: (a) For any k, we have $\emptyset \subsetneq X \subsetneq X \cup X'$ and $X, X \cup X'$ are k-invariant, therefore X is a k-separation of $X \cup X'$ □

Theorem 6.8 We suppose that $X \subset \mathbf{B}^n, X \neq \emptyset$ is k-invariant, $k \in \{(6.1), (6.2)\}$:

(a) if $card(X) = 1$, then X is k-connected;

(b) if $card(X) \geq 2$ and $\mu \in X$ exists with $\Phi(\mu) = \mu$, then $\{\mu\}$ is a k-separation of X;

(c) if $card(X) \geq 2$ and X is k-connected, then it contains no fixed points of Φ.

Proof: We fix an arbitrary k.

(b) As $\Phi(\mu) = \mu$ implies that $\{\mu\}$ is k-invariant and $\emptyset \subsetneq \{\mu\} \subsetneq X$, (b) follows.

(c) If X would have, against all reason, a fixed point μ of Φ, then $\{\mu\}$ would be a k-separation of X, contradiction with the hypothesis that X is k-connected □

Theorem 6.9 Let the functions $\Phi, \Psi : \mathbf{B}^n \longrightarrow \mathbf{B}^n$, the set $X \subset \mathbf{B}^n$ and $(h, h') \in Iso(\Phi, \Psi)$.

(a) We suppose that (6.16) is true, i.e. X is (6.1)-separated by Y. Then

$$\begin{cases} \forall \nu \in \mathbf{B}^n, \Psi^\nu(h(X)) \subset h(X), \\ \exists Y, \emptyset \subsetneq Y \subsetneq h(X) \text{ and } \forall \nu \in \mathbf{B}^n, \Psi^\nu(Y) \subset Y, \end{cases} \tag{6.21}$$

i.e. $h(X)$ is (6.1)-separated by Y.

(b) If (6.17) is true, then

$$\begin{cases} \forall \nu \in \mathbf{B}^n, \Psi^\nu(h(X)) = h(X), \\ \exists Y, \emptyset \subsetneq Y \subsetneq h(X) \text{ and } \forall \nu \in \mathbf{B}^n, \Psi^\nu(Y) = Y. \end{cases} \tag{6.22}$$

Proof: (a) The (6.1)-invariance of X relative to Φ implies from Theorem 5.7, page 60 that $h(X)$ is (6.1)-invariant relative to Ψ. Moreover, the hypothesis (6.16) states the existence of a (6.1)-separation Y' of X relative to Φ. The bijectivity of h shows that $\emptyset \subsetneq h(Y') \subsetneq h(X)$, and Theorem 5.7 shows that $h(Y')$ is (6.1)-invariant relative to Ψ, thus it is a (6.1)-separation of $h(X)$ relative to Ψ.

(b) The proof is similar with the proof of (a) where Theorem 5.7 is replaced by Theorem 5.8, page 60 □

Theorem 6.10 The antiisomorphism $(h, h')^\frown \in Iso^\frown(\Phi, \Psi)$ is given and X nonempty, $X \subset \mathbf{B}^n$. If (6.17) holds, we infer the truth of (6.22).

Proof: This follows from Theorem 5.9, page 64 □

7

Path Connected Set

Under the influence of graph theory, we consider paths[1] as representing finite sequences of distinct points $\mu^0, \mu^1, \dots, \mu^k \in \mathbf{B}^n$ with the property that μ^{i+1} is an immediate successor of μ^i for each $i \in \{0, 1, \dots, k-1\}$. By definition $X \subset \mathbf{B}^n$ nonempty is path connected if for any $\mu \in X, \mu' \in X$ a path exists from μ to μ'.

X is path connected if and only if any $A \subset X$ nonempty is path connected.

If $X_1, \dots, X_k$ is a partition of path connected sets of X, these sets are called path connected components of X.

The morphisms and the antimorphisms bring path connected sets to path connected sets.

7.1 Definition

Definition 7.1 Let $\Phi : \mathbf{B}^n \longrightarrow \mathbf{B}^n$ and $X \subset \mathbf{B}^n$ nonempty. We suppose that the points $\mu, \mu' \in X$ are related by the equation $\mu' = (\Phi^{\nu} \circ \dots \circ \Phi^{\nu'})(\mu)$, where $\nu, \dots, \nu' \in \mathbf{B}^n$ (finitely many such tuples). A **path from** μ **to** μ' is by definition: the point μ, if $\mu = \mu'$, respectively, the finite sequence $\mu, \Phi^{\nu'}(\mu), \dots, (\Phi^{\nu} \circ \dots \circ \Phi^{\nu'})(\mu)$ of distinct values otherwise.

Remark 7.1 We note that several paths may exist from μ to μ'. On the other hand, in the equation $\mu' = (\Phi^{\nu} \circ \dots \circ \Phi^{\nu'})(\mu)$ written for $\mu \neq \mu'$ all of $\nu, \dots, \nu'$ should be chosen non null in order that $\mu, \Phi^{\nu}(\mu), \dots, (\Phi^{\nu} \circ \dots \circ \Phi^{\nu'})(\mu)$ are distinct.

1 In a directed graph, a directed path is a sequence of edges that connect a sequence of vertices, with the added restriction that the edges all be directed in the same direction.

Boolean Functions: Topics in Asynchronicity, First Edition. Serban E. Vlad.

Theorem 7.1 For $X \subset \mathbf{B}^n, X \neq \emptyset$, the properties

$$\forall \mu \in X, X \subset O^+(\mu), \tag{7.1}$$
$$\forall \mu \in X, \forall \mu' \in X, \exists \nu \in \mathbf{B}^n, \ldots, \exists \nu' \in \mathbf{B}^n, \mu' = (\Phi^{\nu} \circ \ldots \circ \Phi^{\nu'})(\mu) \tag{7.2}$$

are equivalent.

Proof: (7.1)$\Longrightarrow$(7.2) Let $\mu \in X, \mu' \in X$ arbitrary. The hypothesis states that $\mu' \in O^+(\mu)$, i.e. $\exists \nu \in \mathbf{B}^n, \ldots, \exists \nu' \in \mathbf{B}^n, \mu' = (\Phi^{\nu} \circ \ldots \circ \Phi^{\nu'})(\mu)$, etc □

Definition 7.2 If one of (7.1), (7.2) is true, we say that X is **path connected.**

Remark 7.2 Intuitively, X is path connected if for any $\mu \in X, \mu' \in X$, we have that μ' is accessible from μ, i.e. a path exists, from μ to μ'. We underline the symmetry of the previous definition, in the sense that if μ' is accessible from μ, then μ is accessible from μ' also.

Remark 7.3 In Definitions 7.1 and 7.2, no request of invariance of X has been made, i.e. some intermediate points of a path from μ to μ' may not belong to X. The option of defining paths and path connectedness like this intends to relate the conjunction of invariance and path connectedness with the attractors, which are the topic of the next chapter.

Theorem 7.2 X_{Φ} is path connected$\Longleftrightarrow \overline{X}_{\Phi^*} = \{\overline{\mu} | \mu \in X\}$ is path connected. In the previous notations, the inferior indexes Φ, Φ^* indicate the functions that path connectedness refers to.

Proof: We refer to (7.2) and we prove $\Longrightarrow$. We take $\mu, \mu' \in \overline{X}$ arbitrary, therefore $\overline{\mu}, \overline{\mu'} \in X$. The hypothesis states the existence of $\nu \in \mathbf{B}^n, \ldots, \nu' \in \mathbf{B}^n$ such that $\overline{\mu'} = (\Phi^{\nu} \circ \ldots \circ \Phi^{\nu'})(\overline{\mu})$, wherefrom

$$\mu' = \overline{(\Phi^{\nu} \circ \ldots \circ \Phi^{\nu'})(\overline{\mu})} = (\Phi^{\nu} \circ \ldots \circ \Phi^{\nu'})^*(\mu)$$
$$\overset{\text{Theorem 1.3, page 8}}{=} (\Phi^{*\nu} \circ \ldots \circ \Phi^{*\nu'})(\mu).$$

The implication $\Longleftarrow$ is proved similarly □

7.2 Examples

Example 7.1 The function from Figure 7.1 fulfills the property that $\forall \mu \in \mathbf{B}^2$, $O^+(\mu) = \mathbf{B}^2$. In such circumstances, any nonempty subset $X \subset \mathbf{B}^2$ is path connected.

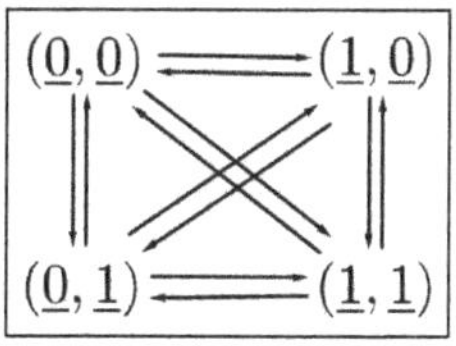

Figure 7.1 The nonempty subsets $X \subset \mathbf{B}^2$ are path connected.

Example 7.2 In Figure 7.2, we have a function with $\forall \mu \in \{(0,0),(0,1)\}$, $O^+(\mu) = \{(0,0),(0,1)\}$ and $\forall \mu \in \{(1,0),(1,1)\}, O^+(\mu) = \{(1,0),(1,1)\}$. We infer in this case that any nonempty $X \subset \{(0,0),(0,1)\}$ and any nonempty $X \subset \{(1,0),(1,1)\}$ is path connected.

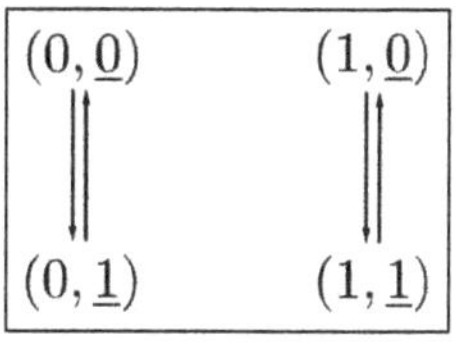

Figure 7.2 The nonempty subsets $X \subset \{(0,0),(0,1)\}$ and $X \subset \{(1,0),(1,1)\}$ are path connected.

Example 7.3 We have in Figure 7.3 a function for which $O^+(0,0) = O^+(1,1) = \mathbf{B}^2$. A nonempty set $X \subset \{(0,0),(1,1)\}$ fulfills $\forall \mu \in X, X \subset \mathbf{B}^2$, i.e. X is path connected.

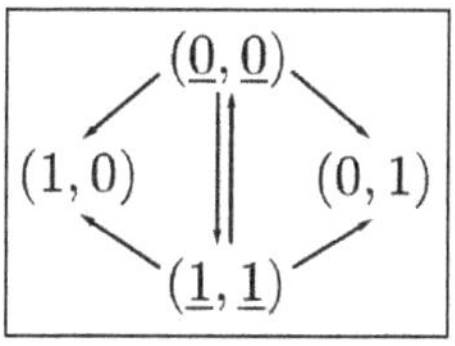

Figure 7.3 The nonempty subsets $X \subset \{(0,0),(1,1)\}$ are path connected.

Example 7.4 In Figure 7.4, we have $O^+(0,0) = O^+(1,1) = O^+(0,1) = \mathbf{B}^2$, and any nonempty set $X \subset \{(0,0),(1,1),(0,1)\}$ is path connected.

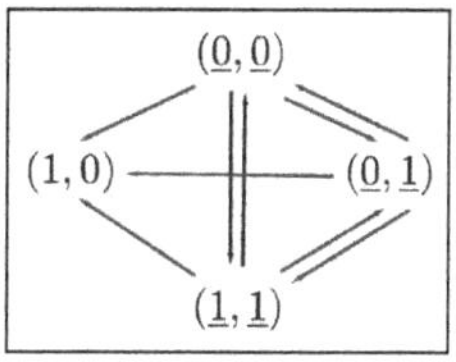

Figure 7.4 The nonempty subsets $X \subset \{(0,0),(1,1),(0,1)\}$ are path connected.

Example 7.5 In Figures 3.3 and 3.4, page 40, the set $\mathbf{B}^2$ is path connected. Any nonempty subset X of $\mathbf{B}^2$ is path connected also, even if some paths have intermediate points that do not belong to X.

7.3 Properties

Theorem 7.3 Let $X \subset \mathbf{B}^n, X \neq \emptyset$.

(a) If $card(X) = 1$, then X is path connected;

(b) if $card(X) \geq 2$ and X contains a fixed point of Φ, then it is not path connected;

(c) we suppose that $card(X) \geq 2$ and X is path connected; then X does not contain a fixed point of Φ.

Proof: (a) For $X = \{\mu\}$, (7.1) takes the form

$$\{\mu\} \subset O^+(\mu).$$

(b) We suppose against all reason that $\mu, \mu' \in X$ exist with $\mu \neq \mu'$ and $\Phi(\mu) = \mu$. We obtain

$$\mu' \notin \{\mu\} = O^+(\mu),$$

contradiction with (7.1).

(c) The supposition against all reason that X contains a fixed point of Φ contradicts (7.1), like at (b) □

Theorem 7.4 Let $\emptyset \subsetneq X \subset \mathbf{B}^n$.

(a) If X is path connected, then any $A \subset X, A \neq \emptyset$ is path connected, in particular a nonempty intersection of path connected sets is path connected.

(b) If any $A \subset X, A \neq \emptyset$ is path connected, then X is path connected.

Proof: (a) If $A \subset X$ is nonempty, then for any $\mu \in A$ we have $A \subset X \subset O^+(\mu)$, thus A is path connected □

Theorem 7.5 If $X_1, X_2 \subset \mathbf{B}^n$ are path connected and $X_1 \cap X_2 \neq \emptyset$, then $X_1 \cup X_2$ is path connected.

Proof: We take $\mu, \mu' \in X_1 \cup X_2$ arbitrary. If $\mu, \mu' \in X_1$, then $\nu, \dots, \nu' \in \mathbf{B}^n$ obviously exist such that $\mu' = (\Phi^{\nu} \circ \dots \circ \Phi^{\nu'})(\mu)$ and the situation is similar for $\mu, \mu' \in X_2$. We suppose now that $\mu \in X_1, \mu' \in X_2$ and let $\mu'' \in X_1 \cap X_2$ arbitrary. From the path connectedness of X_1 and X_2, we have the existence of $\nu, \dots, \nu', \omega, \dots, \omega' \in \mathbf{B}^n$ with $\mu'' = (\Phi^{\nu} \circ \dots \circ \Phi^{\nu'})(\mu)$, $\mu' = (\Phi^{\omega} \circ \dots \circ \Phi^{\omega'})(\mu'')$. We conclude that $\mu' = (\Phi^{\omega} \circ \dots \circ \Phi^{\omega'} \circ \Phi^{\nu} \circ \dots \circ \Phi^{\nu'})(\mu)$ □

7.4 Path Connected Components

Definition 7.3 We consider the nonempty set $X \subset \mathbf{B}^n$ and the partition $X_1, \dots, X_k \subset X$ of path connected subsets of X, $k \geq 2$. Then $X_1, \dots, X_k$ are called **path connected components** of X.

Remark 7.4 Note that in Definition 7.3, X is not necessarily path connected and, on the other hand, that the partition $X_1, \ldots, X_k$ is not unique.

Example 7.6 We have in Figure 7.5, the two sets $X_1 = \{(0,0,0),(0,0,1),(0,1,0),(0,1,1)\}$ and $X_2 = \{(1,0,0),(1,0,1),(1,1,0),(1,1,1)\}$, which are path connected components of $\mathbf{B}^3$. We can partition X_1 and X_2 and get new path connected components of $\mathbf{B}^3$.

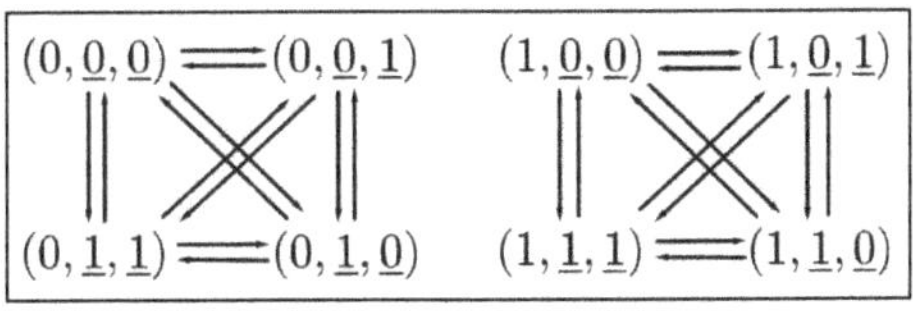

Figure 7.5 The sets $X_1 = \{(0,0,0),(0,0,1),(0,1,0),(0,1,1)\}$, $X_2 = \{(1,0,0),(1,0,1),(1,1,0),(1,1,1)\}$ are path connected components of $\mathbf{B}^3$.

7.5 Morphisms vs Path Connectedness

Theorem 7.6 Let $\Phi, \Psi : \mathbf{B}^n \longrightarrow \mathbf{B}^n$ and $X \subset \mathbf{B}^n$ nonempty. If X_Φ is path connected and $(h, h') \in Hom(\Phi, \Psi)$, then $h(X)_\Psi$ is path connected.

Proof: We take $\lambda, \lambda' \in h(X)$ arbitrary, thus $\mu, \mu' \in X$ exist with $h(\mu) = \lambda, h(\mu') = \lambda'$. Then $\nu \in \mathbf{B}^n, \ldots, \nu' \in \mathbf{B}^n$ exist with $(\Phi^\nu \circ \ldots \circ \Phi^{\nu'})(\mu) = \mu'$, wherefrom

$$\begin{aligned}\lambda' &= h(\mu') = h((\Phi^\nu \circ \ldots \circ \Phi^{\nu'})(\mu)) = (h \circ \Phi^\nu \circ \ldots \circ \Phi^{\nu'})(\mu)\\ &= (\Psi^{h'(\nu)} \circ h \circ \ldots \circ \Phi^{\nu'})(\mu) = \ldots = (\Psi^{h'(\nu)} \circ \ldots \circ h \circ \Phi^{\nu'})(\mu)\\ &= (\Psi^{h'(\nu)} \circ \ldots \circ \Psi^{h'(\nu')} \circ h)(\mu) = (\Psi^{h'(\nu)} \circ \ldots \circ \Psi^{h'(\nu')})(h(\mu))\\ &= (\Psi^{h'(\nu)} \circ \ldots \circ \Psi^{h'(\nu')})(\lambda)\end{aligned}$$

□

7.6 Antimorphisms vs Path Connectedness

Theorem 7.7 If X_Φ is path connected and $(h, h')^\frown \in Hom^\frown(\Phi, \Psi)$, then $h(X)_\Psi$ is path connected.

Proof: We take some arbitrary $\lambda, \lambda' \in h(X)$. This gives the existence of $\mu, \mu' \in X$ such that $h(\mu') = \lambda$ and $h(\mu) = \lambda'$. The path connectedness of X_Φ implies the existence of $\nu \in \mathbf{B}^n, \ldots, \nu' \in \mathbf{B}^n$ with $(\Phi^\nu \circ \ldots \circ \Phi^{\nu'})(\mu) = \mu'$, and we get:

$$\begin{aligned}\lambda' &= h(\mu) = (\Psi^{h'(\nu')} \circ h \circ \Phi^{\nu'})(\mu) = \ldots\\ &= (\Psi^{h'(\nu')} \circ \ldots \circ \Psi^{h'(\nu)} \circ h \circ \Phi^\nu \circ \ldots \circ \Phi^{\nu'})(\mu)\\ &= (\Psi^{h'(\nu')} \circ \ldots \circ \Psi^{h'(\nu)})(h((\Phi^\nu \circ \ldots \circ \Phi^{\nu'})(\mu)))\\ &= (\Psi^{h'(\nu')} \circ \ldots \circ \Psi^{h'(\nu)})(h(\mu')) = (\Psi^{h'(\nu')} \circ \ldots \circ \Psi^{h'(\nu)})(\lambda)\end{aligned}$$

□

8

Attractors

In [16], the attractors X are defined (with different notations) by Andrew Ilachinski in a real space, real time, synchronous context with the following words: 'Although there is no universally accepted definition of an attractor, it is intuitively reasonable to demand that it satisfies the following three properties: (i) Invariance, X is invariant under the map $\Phi : \Phi(X) = X$; (ii) Attractivity, there is an open neighborhood U containing X such that all points $x(t) \in U \longrightarrow X$ as $t \longrightarrow \infty$. The set of initial points $x_i^*(t = 0)$ such that $x_i^*(t)$ approaches X is called the basin of attraction of X, (iii) Irreducibility, X cannot be partitioned into two nonoverlapping invariant and attracting pieces. A more technical demand is that of topological transitivity: there must exist a point $x^* \in X$ such that for all $x \in X$ there exists a positive time T such that $x^*(T)$ is arbitrarily close to x' (the author cites David Ruelle, Floris Takens, Jean-Pierre Eckmann, and Robert Devaney in this definition). In a Boolean asynchronous frame, attractivity is trivial since we can take $U = X$[1] and invariance refers to the strong version $\forall \nu \in \mathbf{B}^n, \Phi^\nu(X) = X$. Adding to invariance either of: minimality (which is (iii) from above), topological transitivity and path connectedness proves to bring equivalent definitions of attractors.

In [23], John Milnor refers (his notations are different from ours) to real space, discrete time synchronous dynamical systems, consisting of a locally compact metric space together with a function Φ from the metric space to itself which describes the evolution of the system in one time step. He mentions that there are completely analogous definitions for systems with continuous time, which are usually defined by autonomous differential equations: 'Let H be a compact set such that $\Phi(H)$ is contained in the interior of H. Then the intersection X of the nested sequence of sets $H \supset \Phi(H) \supset (\Phi \circ \Phi)(H) \supset \ldots$ will be called a trapped attractor, with H as trapping neighborhood' (or trapped attracting set)…'This intersection is always invariant, $\Phi(X) = X$.' We have adopted so far the Boolean weak invariance request $\forall \nu \in \mathbf{B}^n, \Phi^\nu(H) \subset H$

1 In the discrete topology of $\mathbf{B}^n$, all its subsets are open and closed at the same time.

Boolean Functions: Topics in Asynchronicity, First Edition. Serban E. Vlad.

and $X = H \cap \bigcap_{\nu \in \mathbf{B}^n} \Phi^{\nu}(H) \cap \ldots \cap \bigcap_{\nu \in \mathbf{B}^n, \ldots, \nu' \in \mathbf{B}^n} (\Phi^{\nu} \circ \ldots \circ \Phi^{\nu'})(H) \cap \ldots$ and we shall prove at Theorem 8.1 that $\forall \nu \in \mathbf{B}^n, \Phi^{\nu}(X) = X$. Milnor continues: 'The word attractor is usually reserved for an attracting set that contains a dense orbit. (This condition insures that it is not just the union of smaller attracting sets.)' Things are similar with the previous irreducibility of Ilachinski and with our demand of minimality.

Several other authors [1, 3, 13, 17] define the attractors in a manner that proves to be equivalent to Milnor's definition.

Such suggestions that we have grouped around the ideas of Ilachinski and Milnor bring us a unique concept of attractor and its study continues the previous research on invariance and path connectedness. We prove that any $X \subset \mathbf{B}^n, X \neq \emptyset$ with $\forall \nu \in \mathbf{B}^n, \Phi^{\nu}(X) = X$ accepts a partition of attractors.

8.1 Preliminaries

Theorem 8.1 The function $\Phi : \mathbf{B}^n \longrightarrow \mathbf{B}^n$ and the set $H \subset \mathbf{B}^n, H \neq \emptyset$ are considered. We suppose that

$$\forall \nu \in \mathbf{B}^n, \Phi^{\nu}(H) \subset H, \tag{8.1}$$

and we define the sequence of sets:

$$\begin{aligned} X_0 &= H, \\ X_1 &= \bigcap_{\nu \in \mathbf{B}^n} \Phi^{\nu}(H), \\ &\vdots \\ X_k &= \bigcap_{\nu^1 \in \mathbf{B}^n} \ldots \bigcap_{\nu^k \in \mathbf{B}^n} (\Phi^{\nu^1} \circ \ldots \circ \Phi^{\nu^k})(H), \\ &\vdots \end{aligned} \tag{8.2}$$

X_k is descending and convergent toward a limit $X = \lim_{k \to \infty} X_k$. If $X \neq \emptyset$, we have

$$\forall \nu \in \mathbf{B}^n, \Phi^{\nu}(X) = X. \tag{8.3}$$

Proof: We have indeed

$$\bigcap_{\nu \in \mathbf{B}^n} \Phi^{\nu}(H) \overset{(8.1)}{\subset} H, \tag{8.4}$$

thus $X_1 \subset X_0$. Moreover,

$$\begin{aligned} X_2 &= \bigcap_{\nu^1 \in \mathbf{B}^n} \bigcap_{\nu^2 \in \mathbf{B}^n} (\Phi^{\nu^1} \circ \Phi^{\nu^2})(H) \\ &= \bigcap_{\nu^1 \in \mathbf{B}^n} \Phi^{\nu^1}\Big(\bigcap_{\nu^2 \in \mathbf{B}^n} \Phi^{\nu^2}(H)\Big) \overset{(8.4)}{\subset} \bigcap_{\nu^1 \in \mathbf{B}^n} \Phi^{\nu^1}(H) = X_1 \end{aligned}$$

and the fact that X_k is descending is proved by induction on k.

Let now $k_1 \in \mathbf{N}$ with the property that $\forall k \geq k_1, X_k = X_{k_1}$ and the hypothesis states that $X = X_{k_1}$ is nonempty. We can write that

$$X_{k_1+1} = \bigcap_{\nu^1 \in \mathbf{B}^n} \bigcap_{\nu^2 \in \mathbf{B}^n} \dots \bigcap_{\nu^{k_1+1} \in \mathbf{B}^n} (\Phi^{\nu^1} \circ \Phi^{\nu^2} \circ \dots \circ \Phi^{\nu^{k_1+1}})(H)$$

$$= \bigcap_{\nu^1 \in \mathbf{B}^n} \Phi^{\nu^1}(\bigcap_{\nu^2 \in \mathbf{B}^n} \dots \bigcap_{\nu^{k_1+1} \in \mathbf{B}^n} (\Phi^{\nu^2} \circ \dots \circ \Phi^{\nu^{k_1+1}})(H)) = \bigcap_{\nu^1 \in \mathbf{B}^n} \Phi^{\nu^1}(X_{k_1}),$$

therefore

$$\bigcap_{\nu^1 \in \mathbf{B}^n} \Phi^{\nu^1}(X_{k_1}) = X_{k_1}. \tag{8.5}$$

Equation (8.5) is equivalent with (8.3) □

Theorem 8.2 Let $X \subset \mathbf{B}^n$ nonempty. The following properties are equivalent:
(a) H exists, $X \subset H \subset \mathbf{B}^n$ such that the invariance (8.1) holds and the sequence X_k defined at (8.2) has the limit X,
(b) X satisfies the invariance property (8.3).

Proof: (a) $\Longrightarrow$ (b) We suppose the existence of H such that $X \subset H \subset \mathbf{B}^n$ and (8.1) is true. The fact that X_k that are defined by (8.2) have the limit X implies (8.3) was proved at Theorem 8.1.

(b) $\Longrightarrow$ (a) If X satisfies (8.3), then the sequence (8.2) defined for $H = X$ is constant $X_k = X$ and has the limit X. (8.3) implies also (8.1) □

Remark 8.1 The previous reasoning allows thinking of attractors as defined by Milnor, translated in a Boolean context, as sets X fulfilling (8.3) and also minimality. This gives the same point of view on attractors like Ilachinski's.

8.2 Definition

Definition 8.1 We consider the function $\Phi : \mathbf{B}^n \longrightarrow \mathbf{B}^n$ and the nonempty set $X \subset \mathbf{B}^n$. If X fulfills one of

$$\begin{cases} \forall \nu \in \mathbf{B}^n, \Phi^\nu(X) = X, \\ \exists \mu^* \in X, \forall \mu \in X, \exists \nu \in \mathbf{B}^n, \dots, \exists \nu' \in \mathbf{B}^n, (\Phi^\nu \circ \dots \circ \Phi^{\nu'})(\mu^*) = \mu, \end{cases} \tag{8.6}$$

$$\begin{cases} \forall \nu \in \mathbf{B}^n, \Phi^\nu(X) = X, \\ \forall \mu \in X, \forall \mu' \in X, \exists \nu \in \mathbf{B}^n, \dots, \exists \nu' \in \mathbf{B}^n, (\Phi^\nu \circ \dots \circ \Phi^{\nu'})(\mu) = \mu', \end{cases} \tag{8.7}$$

$$\begin{cases} \forall \nu \in \mathbf{B}^n, \Phi^\nu(X) = X, \\ \forall Y, (\emptyset \subsetneq Y \subset X \text{ and } \forall \nu \in \mathbf{B}^n, \Phi^\nu(Y) = Y) \Longrightarrow X = Y, \end{cases} \tag{8.8}$$

it is called **attractor**.

Remark 8.2 The concept of attractor has many definitions in literature and for this reason its binary translation can be made in different ways. We shall prove later, in Theorem 8.6, that (8.6)–(8.8) are equivalent. We notice for the moment that

- in (8.6) we have topological transitivity;
- in (8.7) we have path connectedness;
- in (8.8) we have minimality (connectedness).

Example 8.1 We have drawn in Figure 8.1 the state portrait of the function $\Phi : \mathbf{B}^3 \longrightarrow \mathbf{B}^3$, $\Phi(\mu_1, \mu_2, \mu_3) = (\mu_1 \oplus \mu_3, \mu_2, \mu_2 \oplus \mu_3)$. $X = \{(0,1,0), (0,1,1), (1,1,1), (1,1,0)\}$ fulfills all of (8.6), ..., (8.8).

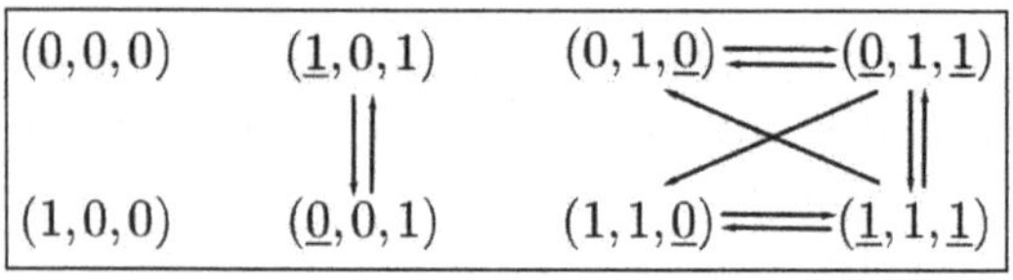

Figure 8.1 The set $X = \{(0,1,0), (0,1,1), (1,1,1), (1,1,0)\}$ is an attractor.

8.3 Properties

Theorem 8.3 Let $X \subset \mathbf{B}^n$ nonempty and we ask that

$$\forall \nu \in \mathbf{B}^n, \Phi^\nu(X) = X.$$

The following statements are equivalent:

(a) X is disconnected: $\exists X_1$,

$$\emptyset \subsetneq X_1 \subsetneq X \text{and } \forall \nu \in \mathbf{B}^n, \Phi^\nu(X_1) = X_1, \tag{8.9}$$

(b) $X_1 \subset X, X_2 \subset X$ nonempty exist with $X_1 \cap X_2 = \emptyset, X_1 \cup X_2 = X$ and X_i are invariant:

$$\forall i \in \{1,2\}, \forall \nu \in \mathbf{B}^n, \Phi^\nu(X_i) = X_i.$$

Proof: (a) $\Longrightarrow$ (b) Let X_1 a separation of X, thus (8.9) is true. As $X \backslash X_1 \neq \emptyset$, studying its invariance makes sense. Let us suppose that (b) is false, namely that

$$\exists \nu \in \mathbf{B}^n, \Phi^\nu(X \backslash X_1) \neq X \backslash X_1. \tag{8.10}$$

As Φ^ν restricted to X is bijective, we infer that Φ^ν restricted to $X \backslash X_1$ is injective and (8.10) implies

$$\exists \mu \in X, \mu \notin X_1, \Phi^\nu(\mu) \in X_1.$$

From the invariance of X_1, we infer that $\mu = (\Phi^\nu)^{-1}(\Phi^\nu(\mu)) \in X_1$, contradiction. The statement of the theorem is true for $X_2 = X \backslash X_1$.

(b) $\Longrightarrow$ (a) X_1 is a separation of X, thus X is disconnected □

Remark 8.3 Theorem 8.3 cannot be stated for the $\forall \nu \in \mathbf{B}^n, \Phi^\nu(X) \subset X$ invariance of X : note that in the proof of (a)$\Longrightarrow$(b) we have used the bijectivity of Φ^ν restricted to X.

Theorem 8.4 Let $X \subset \mathbf{B}^n$ nonempty such that $\forall \nu \in \mathbf{B}^n$,

$$\Phi^\nu(X) = X \tag{8.11}$$

is true. Then

(a) $\forall \nu \in \mathbf{B}^n, \forall \mu \in X$, the sequence $\mu, (\Phi^\nu)(\mu), (\Phi^\nu)^{(2)}(\mu), \dots$ is periodic.

(b) $p \geq 1$ exists such that $\forall \nu \in \mathbf{B}^n, \forall \mu \in X$,

$$(\Phi^\nu)^{(p)}(\mu) = \mu, \tag{8.12}$$

i.e. if we denote abusively with the same symbol Φ^ν the restriction of this function to X, we have $(\Phi^\nu)^{-1} = (\Phi^\nu)^{(p-1)}$.

Proof: (a) We take $\nu \in \mathbf{B}^n, \mu \in X$ arbitrary and fixed. As X is a finite set, the values of the sequence $\mu, (\Phi^\nu)(\mu), (\Phi^\nu)^{(2)}(\mu), \dots$ are finitely many, they repeat. We denote with μ' the first value that repeats and we claim that $\mu = \mu'$. If this would not be true, then we would have the existence of $p \geq 1, p' \geq 1$ with

$$\mu' = (\Phi^\nu)^{(p)}(\mu) = (\Phi^\nu)^{(p+p')}(\mu),$$

meaning that

$$(\Phi^\nu)^{(p-1)}(\mu) \neq (\Phi^\nu)^{(p+p'-1)}(\mu)$$

and

$$\Phi^\nu((\Phi^\nu)^{(p-1)}(\mu)) = \Phi^\nu((\Phi^\nu)^{(p+p'-1)}(\mu))$$

are true, representing a contradiction with the bijectivity of Φ^ν restricted to X.

It has resulted that $\mu = \mu'$ and $p \geq 1$ exists with $\mu = (\Phi^\nu)^{(p)}(\mu)$. The sequence from the statement of the theorem is periodic with the period p.

(b) Let $\nu \in \mathbf{B}^n, \mu \in X$ arbitrary for which the sequence μ, $(\Phi^\nu)(\mu)$, $(\Phi^\nu)^{(2)}(\mu), \ldots$ is periodic from item (a) and we denote with $p_{\nu,\mu}$ its least period. The set $P = \{p_{\nu,\mu} | \nu \in \mathbf{B}^n, \mu \in X\}$ is finite and we use the notation p for the least common multiple of $p_{\nu,\mu} \in P$. p satisfies equation (8.12) for all $\nu \in \mathbf{B}^n$ and $\mu \in X$ □

Theorem 8.5 We suppose that $\forall \nu \in \mathbf{B}^n$, (8.11) holds and we take an arbitrary $\mu \in X$. The set $O^+(\mu)$ satisfies:

(a) $\forall \nu \in \mathbf{B}^n$,

$$\Phi^\nu(O^+(\mu)) = O^+(\mu),$$

(b) it is path connected:

$$\forall \mu' \in O^+(\mu), \forall \mu'' \in O^+(\mu), \exists \nu \in \mathbf{B}^n, \ldots, \exists \nu' \in \mathbf{B}^n,$$
$$\mu'' = (\Phi^\nu \circ \ldots \circ \Phi^{\nu'})(\mu').$$

Proof: (a) Let $\nu \in \mathbf{B}^n$ arbitrary, fixed. We prove

$$\Phi^\nu(O^+(\mu)) \subset O^+(\mu)$$

and we take for this $\mu' \in O^+(\mu)$ arbitrary. We infer the existence of $\nu' \in \mathbf{B}^n, \ldots, \nu'' \in \mathbf{B}^n$ with $\mu' = (\Phi^{\nu'} \circ \ldots \circ \Phi^{\nu''})(\mu)$. Obviously $(\Phi^\nu \circ \Phi^{\nu'} \circ \ldots \circ \Phi^{\nu''})(\mu) \in O^+(\mu)$. We prove now

$$O^+(\mu) \subset \Phi^\nu(O^+(\mu))$$

and we take for this $\mu' = (\Phi^{\nu'} \circ \ldots \circ \Phi^{\nu''})(\mu) \in O^+(\mu)$ arbitrary, where $\nu' \in \mathbf{B}^n, \ldots, \nu'' \in \mathbf{B}^n$. From Theorem 8.4, we get the existence of $p \geq 1$ with the property that $\forall \mu \in X$,

$$(\Phi^\nu)^{(p)}(\mu) = \mu.$$

We have

$$\mu' = (\Phi^{\nu'} \circ \ldots \circ \Phi^{\nu''})(\mu) = (\Phi^\nu \circ (\Phi^\nu)^{-1})(\Phi^{\nu'} \circ \ldots \circ \Phi^{\nu''})(\mu)$$
$$= (\Phi^\nu \circ (\Phi^\nu)^{(p-1)} \circ \Phi^{\nu'} \circ \ldots \circ \Phi^{\nu''})(\mu) \in \Phi^\nu(O^+(\mu)).$$

(b) We take $\mu' \in O^+(\mu)$, $\mu'' \in O^+(\mu)$ arbitrary, fixed. From the way that $O^+(\mu)$ was defined, $\nu, \ldots, \nu', \omega, \ldots, \omega' \in \mathbf{B}^n$ exist such that

$$\mu' = (\Phi^\nu \circ \ldots \circ \Phi^{\nu'})(\mu),$$
$$\mu'' = (\Phi^\omega \circ \ldots \circ \Phi^{\omega'})(\mu).$$

We infer that

$$\mu'' = ((\Phi^\omega \circ \ldots \circ \Phi^{\omega'}) \circ (\Phi^\nu \circ \ldots \circ \Phi^{\nu'})^{-1})(\mu')$$
$$= (\Phi^\omega \circ \ldots \circ \Phi^{\omega'} \circ (\Phi^{\nu'})^{-1} \circ \ldots \circ (\Phi^\nu)^{-1})(\mu').$$

From Theorem 8.4, we have the existence of $p \geq 1$ with $(\Phi^{\nu'})^{-1} = (\Phi^{\nu'})^{(p-1)}, \dots,$ $(\Phi^{\nu})^{-1} = (\Phi^{\nu})^{(p-1)}$, therefore

$$\mu'' = (\Phi^{\omega} \circ \dots \circ \Phi^{\omega'} \circ (\Phi^{\nu'})^{(p-1)} \circ \dots \circ (\Phi^{\nu})^{(p-1)})(\mu') \qquad \square$$

Theorem 8.6 We consider the function $\Phi : \mathbf{B}^n \longrightarrow \mathbf{B}^n$. For $X \subset \mathbf{B}^n, X \neq \emptyset$ the statements (8.6), (8.7) and (8.8) are equivalent.

Proof: (8.6) $\Longrightarrow$ (8.7) Let $\mu^* \in X$ with the property that for arbitrary, fixed $\mu \in X$, we have the existence of $\nu, \dots, \nu' \in \mathbf{B}^n$ such that

$$(\Phi^{\nu} \circ \dots \circ \Phi^{\nu'})(\mu^*) = \mu.$$

The restrictions of $\Phi^{\nu}, \dots, \Phi^{\nu'}$ to X are bijective, thus we can write:

$$\mu^* = (\Phi^{\nu} \circ \dots \circ \Phi^{\nu'})^{-1}(\mu) = ((\Phi^{\nu'})^{-1} \circ \dots \circ (\Phi^{\nu})^{-1})(\mu)$$

and $p \geq 1$ exists with the property that $(\Phi^{\nu})^{-1} = (\Phi^{\nu})^{(p-1)}, \dots, (\Phi^{\nu'})^{-1} = (\Phi^{\nu'})^{(p-1)}$. We have

$$\mu^* = ((\Phi^{\nu'})^{(p-1)} \circ \dots \circ (\Phi^{\nu})^{(p-1)})(\mu).$$

We take now an arbitrary $\mu' \in X$, for which $\omega, \dots, \omega' \in \mathbf{B}^n$ exist with

$$\mu' = (\Phi^{\omega} \circ \dots \circ \Phi^{\omega'})(\mu^*).$$

We conclude that

$$\begin{aligned}\mu' &= (\Phi^{\omega} \circ \dots \circ \Phi^{\omega'})(((\Phi^{\nu'})^{(p-1)} \circ \dots \circ (\Phi^{\nu})^{(p-1)})(\mu)) \\ &= (\Phi^{\omega} \circ \dots \circ \Phi^{\omega'} \circ (\Phi^{\nu'})^{(p-1)} \circ \dots \circ (\Phi^{\nu})^{(p-1)})(\mu).\end{aligned}$$

(8.7)$\Longrightarrow$(8.6) Obvious.
(8.7)$\Longrightarrow$(8.8) We suppose against all reason the existence of Y with $\emptyset \subsetneq Y \subsetneq X$ and $\forall \nu \in \mathbf{B}^n, \Phi^{\nu}(Y) = Y$. Theorem 8.3 shows that $X \setminus Y$ is invariant. Let $\mu \in Y$ and $\mu' \in X \setminus Y$, for which we have the existence of $\nu \in \mathbf{B}^n, \dots, \nu' \in \mathbf{B}^n$ with $\mu' = (\Phi^{\nu} \circ \dots \circ \Phi^{\nu'})(\mu)$. This equation together with the invariance of Y gives $\mu' \in Y$, contradiction. Such a μ' does not exist, thus $X = Y$.

(8.8)$\Longrightarrow$(8.7) We suppose the contrary, namely that $\mu, \mu' \in X$ exist such that $\forall \nu \in \mathbf{B}^n, \dots, \forall \nu' \in \mathbf{B}^n$,

$$\mu' \neq (\Phi^{\nu} \circ \dots \circ \Phi^{\nu'})(\mu).$$

But the set

$$O^+(\mu) = \{(\Phi^{\nu} \circ \dots \circ \Phi^{\nu'})(\mu) | \nu \in \mathbf{B}^n, \dots, \nu' \in \mathbf{B}^n\}$$

satisfies $\forall \nu \in \mathbf{B}^n, \Phi^{\nu}(O^+(\mu)) = O^+(\mu)$ from Theorem 8.5, therefore $\mu' \in X \setminus O^+(\mu)$. A contradiction has resulted with the supposition that X is minimal $\square$

Corollary 8.1 Let $X \subset \mathbf{B}^n, X \neq \emptyset$. The fact that $\forall \nu \in \mathbf{B}^n$, the invariance (8.11) holds, implies the existence of the partition $X_1, \dots, X_p$ of X, which is unique modulo the order of these sets, such that

(a) $\forall \nu \in \mathbf{B}^n, \forall i \in \{1, \dots, p\}$,

$$\Phi^\nu(X_i) = X_i,$$

(b) $X_1, \dots, X_p$ are minimal, path connected and topologically transitive.

Proof: We take $\mu \in X$ arbitrary and define $X_1 = O^+(\mu)$. Theorem 8.5 shows that X_1 is invariant and path connected. On the other hand Theorem 8.3 (a)$\Longrightarrow$ (b), page 90 shows that $X \backslash X_1$ is invariant, thus we can take $\mu' \in X \backslash X_1$ arbitrary for which we define $X_2 = O^+(\mu')$. As X_2 is invariant and path connected, we continue the reasoning with $X \backslash (X_1 \cup X_2)$. In p steps we obtain the partition $X_1, \dots, X_p$ of X that satisfies the requirements of invariance and path connectedness. By Theorem 8.6, path connectedness implies minimality and topological transitivity.

The uniqueness of the partition $X_1, \dots, X_p$ follows from Theorem 6.4, page 76 □

8.4 Morphisms vs Attractors

Theorem 8.7 The functions $\Phi, \Psi : \mathbf{B}^n \longrightarrow \mathbf{B}^n$ are considered together with $(h, h') \in Iso(\Phi, \Psi)$ and we take $X \subset \mathbf{B}^n, X \neq \emptyset$. If X is attractor, then $h(X)$ is attractor.

Proof: We suppose that (8.7) is true. The invariance

$$\forall \nu \in \mathbf{B}^n, \Phi^\nu(X) = X \tag{8.13}$$

implies from Theorem 5.8, page 60 the invariance

$$\forall \nu \in \mathbf{B}^n, \Psi^\nu(h(X)) = h(X). \tag{8.14}$$

On the other hand, the path connectedness of X_Φ:

$$\forall \mu \in X, \forall \mu' \in X, \exists \nu \in \mathbf{B}^n, \dots, \exists \nu' \in \mathbf{B}^n, (\Phi^\nu \circ \dots \circ \Phi^{\nu'})(\mu) = \mu'$$

implies from Theorem 7.6, page 85 the path connectedness of $h(X)_\Psi$ □

8.5 Antimorphisms vs Attractors

Theorem 8.8 Let $\Phi, \Psi : \mathbf{B}^n \longrightarrow \mathbf{B}^n$, $(h, h')^{\frown} \in Iso^{\frown}(\Phi, \Psi)$ and $X \subset \mathbf{B}^n$ nonempty. If X is attractor, then $h(X)$ is also attractor.

Proof: If (8.13) is true, then the invariance (8.14) takes place, from Theorem 5.9, page 64. If in addition, X_Φ is path connected, then $h(X)_\Psi$ is path connected, from Theorem 7.7, page 85 □

9

The Technical Condition of Proper Operation

We suppose that the model of a circuit is represented by the function $\Phi : \mathbf{B}^n \longrightarrow \mathbf{B}^n$. The situation when $\mu \in \mathbf{B}^n$ exists such that $card(\{i|i \in \{1,\ldots,n\}, \Phi_i(\mu) \neq \mu_i\}) > 1$ is called a race, the coordinates i,j with $\Phi_i(\mu) \neq \mu_i, \Phi_j(\mu) \neq \mu_j$ are "racing" to see which one can change first and this generates an unpredictable behavior of the circuit, since the speeds of computation of Φ_i and Φ_j are not known. To avoid the races that could occur, Φ is sometimes specified so that for any μ, at most one its coordinates can change when Φ is computed; such a circuit is called race-free and we also say that Φ fulfills the technical condition of proper operation (tcpo).

We have that Φ fulfills tcpo$\Longleftrightarrow \forall \nu \in \mathbf{B}^n, \Phi^\nu$ fulfills tcpo$\Longleftrightarrow \forall \mu \in \mathbf{B}^n, \mu^- = \{\mu\} \cup \Phi^{-1}(\mu) \Longleftrightarrow \forall \mu \in \mathbf{B}^n, \mu^+ = \{\mu\} \cup \{\Phi(\mu)\}$.

Certain special forms of $\Phi^{-1}(\mu)$ and μ^- give also equivalent conditions with the fact that Φ fulfills tcpo.

The sources, the isolated fixed points, the transient points and the sinks are characterized in terms of $\Phi^{-1}(\mu)$ when tcpo is satisfied.

We show that, in certain circumstances, the isomorphisms and the antiisomorphisms bring functions that fulfill tcpo in functions that fulfill tcpo.

9.1 Definition

Theorem 9.1 For $\Phi : \mathbf{B}^n \longrightarrow \mathbf{B}^n$, the following statements (a), (b), (c) are equivalent:

(a) $\forall \mu \in \mathbf{B}^n$,

$$card(\Phi_\mu) \in \{0, 1\}, \tag{9.1}$$

see Definition 1.12 and Remark 1.16, page 12, where Φ_μ is the set of the unstable coordinates of μ;

(b) $\forall \mu \in \mathbf{B}^n$, one of the following properties is true:

$$\Phi(\mu) = \mu, \tag{9.2}$$

$$\exists i \in \{1, \ldots, n\}, \Phi(\mu) = \mu \oplus \varepsilon^i; \tag{9.3}$$

(c) $\forall \mu \in \mathbf{B}^n$, one of the following properties is true:

$$\Phi^{-1}(\mu) = \emptyset, \tag{9.4}$$

$$\Phi^{-1}(\mu) = \{\mu\}, \tag{9.5}$$

$$\exists i \in \{1, \dots, n\}, \Phi^{-1}(\mu) = \{\mu \oplus \varepsilon^i\}, \tag{9.6}$$

$$\exists i \in \{1, \dots, n\}, \Phi^{-1}(\mu) = \{\mu, \mu \oplus \varepsilon^i\}, \tag{9.7}$$

$$\exists i \in \{1, \dots, n\}, \exists j \in \{1, \dots, n\}, \Phi^{-1}(\mu) = \{\mu \oplus \varepsilon^i, \mu \oplus \varepsilon^j\}, \tag{9.8}$$

$$\exists i \in \{1, \dots, n\}, \exists j \in \{1, \dots, n\}, \Phi^{-1}(\mu) = \{\mu, \mu \oplus \varepsilon^i, \mu \oplus \varepsilon^j\}, \tag{9.9}$$

$$\vdots$$

$$\exists i_1 \in \{1, \dots, n\}, \dots, \exists i_n \in \{1, \dots, n\},$$
$$\Phi^{-1}(\mu) = \{\mu \oplus \varepsilon^{i_1}, \dots, \mu \oplus \varepsilon^{i_n}\}, \tag{9.10}$$

$$\exists i_1 \in \{1, \dots, n\}, \dots, \exists i_n \in \{1, \dots, n\},$$
$$\Phi^{-1}(\mu) = \{\mu, \mu \oplus \varepsilon^{i_1}, \dots, \mu \oplus \varepsilon^{i_n}\}. \tag{9.11}$$

Proof: (a)$\Longleftrightarrow$(b) For any $\mu \in \mathbf{B}^n$, $card(\{i | i \in \{1, \dots, n\}, \Phi_i(\mu) \neq \mu_i\}) = 0$ is equivalent with (9.2) (all the coordinates of μ are stable) and $card(\{i | i \in \{1, \dots, n\}, \Phi_i(\mu) \neq \mu_i\}) = 1$ is equivalent with (9.3) (μ has exactly one unstable coordinate).

(b)$\Longrightarrow$(c) Let us fix an arbitrary $\mu \in \mathbf{B}^n$ and we suppose against all reason that (9.4)–(9.11) are all false. This means the existence of $p \in \{2, \dots, n\}$ and $i_1, \dots, i_p \in \{1, \dots, n\}$ distinct such that $\mu \oplus \varepsilon^{i_1} \oplus \dots \oplus \varepsilon^{i_p} \in \Phi^{-1}(\mu)$. Then $\Phi(\mu \oplus \varepsilon^{i_1} \oplus \dots \oplus \varepsilon^{i_p}) = \mu$ and (9.2), (9.3) are both false for $\mu' = \mu \oplus \varepsilon^{i_1} \oplus \dots \oplus \varepsilon^{i_p}$, contradiction.

(c)$\Longrightarrow$(b) We suppose against all reason that (b) is false. This means the existence of $\mu \in \mathbf{B}^n$, $p \in \{2, \dots, n\}$ and $i_1, \dots, i_p \in \{1, \dots, n\}$ distinct such that $\Phi(\mu) = \mu \oplus \varepsilon^{i_1} \oplus \dots \oplus \varepsilon^{i_p}$. We infer that $\mu \in \Phi^{-1}(\mu \oplus \varepsilon^{i_1} \oplus \dots \oplus \varepsilon^{i_p})$, i.e. (9.4)–(9.11) are all false for $\mu' = \mu \oplus \varepsilon^{i_1} \oplus \dots \oplus \varepsilon^{i_p}$, contradiction □

Definition 9.1 The function Φ is said to fulfill the **technical condition of proper operation** (tcpo), or that it is **race-free**, if one of the previous properties (a), (b), (c) holds.

Remark 9.1 Intuitively, tcpo states that for all μ, the n-tuples μ and $\Phi(\mu)$ differ on at most one coordinate. When Φ models a circuit, this is a sufficient property for the predictability of the circuit, taking into account the unknown parameters that occur: for any μ and any order (any speed) of computation of $\Phi_1(\mu), \dots, \Phi_n(\mu)$, we have that $\Phi(\mu)$ is computed.

Theorem 9.2 Φ fulfills tcpo if and only if Φ^* fulfills tcpo.

Proof: We take $\mu \in \mathbf{B}^n$ and $i \in \{1,\dots,n\}$ arbitrary. We have

$$\mu_i \oplus \Phi_i(\mu) = (1 \oplus \mu_i) \oplus (1 \oplus \Phi_i(\mu)) = \overline{\mu_i} \oplus \overline{\Phi_i(\mu)} = \overline{\mu_i} \oplus \overline{\Phi_i(\overline{\overline{\mu}})}$$
$$= \overline{\mu_i} \oplus \Phi_i^*(\overline{\mu}),$$

therefore

$$\Phi_\mu = \{j | j \in \{1,\dots,n\}, \mu_j \oplus \Phi_j(\mu) = 1\}$$
$$= \{j | j \in \{1,\dots,n\}, \overline{\mu_j} \oplus \Phi_j^*(\overline{\mu}) = 1\} = \Phi_{\overline{\mu}}^*.$$

This means that the equivalence

$$\forall \mu \in \mathbf{B}^n, card(\Phi_\mu) \in \{0,1\} \Longleftrightarrow \forall \mu \in \mathbf{B}^n, card(\Phi_{\overline{\mu}}^*) \in \{0,1\}$$

holds □

Theorem 9.3 If Φ is bijective, then the following statements are equivalent:
(a) Φ fulfills tcpo;
(b) $\forall \mu \in \mathbf{B}^n$, (9.5) or (9.6) is true.

Proof: We take $\mu \in \mathbf{B}^n$ arbitrary and fixed.
(a)$\Longrightarrow$(b) The hypothesis states the truth of the disjunction of (9.4)–(9.11).
Case (1), (9.4)
We have two possibilities: $\Phi(\mu) = \mu$, giving the contradiction $\mu \in \Phi^{-1}(\mu)$ with hypothesis (9.4), and $\exists i \in \{1,\dots,n\}, \Phi(\mu) = \mu \oplus \varepsilon^i$. We suppose that the last possibility is fulfilled and we have two other possibilities: $\Phi^{(2)}(\mu) = \Phi(\mu)$, giving the contradiction $\mu, \Phi(\mu) \in \Phi^{-1}(\Phi(\mu))$, and $\exists j \in \{1,\dots,n\}, \Phi^{(2)}(\mu) = \Phi(\mu) \oplus \varepsilon^j$. If the last possibility is fulfilled, we have two new possibilities: $\Phi^{(3)}(\mu) = \Phi^{(2)}(\mu)$, contradiction since $\Phi(\mu), \Phi^{(2)}(\mu) \in \Phi^{-1}(\Phi^{(2)}(\mu))$, and $\exists k \in \{1,\dots,n\}, \Phi^{(3)}(\mu) = \Phi^{(2)}(\mu) \oplus \varepsilon^k$. We continue the reasoning and we obtain a sequence $\mu = \Phi^{(0)}(\mu)$, $\Phi^{(1)}(\mu), \Phi^{(2)}(\mu), \dots$ where $\Phi^{(k+1)}(\mu) = \Phi^{(k)}(\mu) \oplus \varepsilon^{i_k}, k \in \mathbf{N}$, as indicated by the bijectivity of Φ. But this sequence has finitely many distinct values and let $k_1 \geq 2$ be the rank which is defined like this:

$$\Phi^{(0)}(\mu), \Phi^{(1)}(\mu), \dots, \Phi^{(k_1-1)}(\mu) \text{ are distinct,}$$

$$\exists k_2 \in \{0,\dots,k_1-1\}, \Phi^{(k_1)}(\mu) = \Phi^{(k_2)}(\mu).$$

If $k_2 = 0$, thus $\Phi^{(k_1)}(\mu) = \Phi^{(0)}(\mu) = \mu$, then $\Phi^{(k_1-1)}(\mu) \in \Phi^{-1}(\mu)$, representing a contradiction with (9.4), and this implies that $k_2 \geq 1$. In this last situation, we have

$$\Phi^{(k_2-1)}(\mu) \neq \Phi^{(k_1-1)}(\mu),$$

$$\Phi(\Phi^{(k_2-1)}(\mu)) = \Phi^{(k_2)}(\mu) = \Phi^{(k_1)}(\mu) = \Phi(\Phi^{(k_1-1)}(\mu)),$$

contradiction again with the bijectivity of Φ. It has resulted that (9.4) is false.

Case (2), (9.5)
This is true if (9.2) is true.
Case (3), (9.6)
This is true if (9.3) is true, under the form: $\exists i \in \{1, \ldots, n\}, \Phi(\mu \oplus \varepsilon^i) = \mu$.
Case (4), (9.7)
$\mu \neq \mu \oplus \varepsilon^i$ and $\Phi(\mu) = \Phi(\mu \oplus \varepsilon^i)(= \mu)$ is a contradiction;
Case (5), (9.8)
$\mu \oplus \varepsilon^i \neq \mu \oplus \varepsilon^j$ and $\Phi(\mu \oplus \varepsilon^i) = \Phi(\mu \oplus \varepsilon^j)(= \mu)$ is a contradiction;
... At this moment, any of (9.9)–(9.11) gives a contradiction.

(b)$\Longrightarrow$(a) The truth of the disjunction of (9.5), (9.6) implies the truth of the disjunction of (9.4)–(9.11) □

9.2 Examples

Example 9.1 The identity $1_{\mathbf{B}^n} : \mathbf{B}^n \longrightarrow \mathbf{B}^n$ fulfills tcpo, since all $\mu \in \mathbf{B}^n$ are fixed points of $1_{\mathbf{B}^n}$ (equation (9.2) is satisfied).

Example 9.2 In Figure 9.1, tcpo is satisfied under the form: in $(0, 0)$, $(0, 1)$ equation (9.3) is true and in the points $(1, 1)$, $(1, 0)$ equation (9.2) is true.

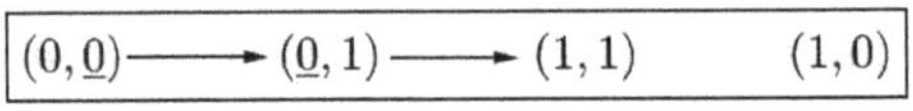

Figure 9.1 Φ fulfills tcpo.

Example 9.3 The function whose state portrait is drawn in Figure 9.2 fulfills tcpo also: in $(1, 0), (1, 1), (0, 1)$ equation (9.2) holds, and in $(0, 0)$ we have the truth of (9.3).

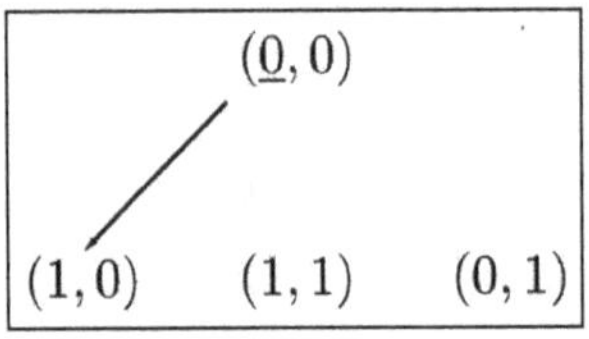

Figure 9.2 Φ fulfills tcpo.

Example 9.4 Let us get back to the function from Figure 5.1, page 57, where tcpo is also true; in all the points (9.3) holds.

Example 9.5 In Figure 5.2, page 57 tcpo holds in the following manner: (9.2) is true in $(0, 0), (0, 1)$ and (9.3) is true in $(1, 0), (1, 1)$.

Example 9.6 In Figure 5.3, page 57, tcpo is true under the form: in all $\mu \in \mathbf{B}^2$, (9.3) is satisfied.

Example 9.7 The functions Φ from Examples 9.1, 9.4, and 9.6 are bijective and we can take a look how Theorem 9.3 works.

9.3 Iterates

Remark 9.2 If $\Phi : \mathbf{B}^n \longrightarrow \mathbf{B}^n$ fulfills tcpo, then $\Phi \circ \Phi$ might not fulfill the same property. We give the example from Figure 9.1, in which $(\Phi \circ \Phi)(0,0) = \Phi(0,1) = (1,1)$. More general, if $\Psi : \mathbf{B}^n \longrightarrow \mathbf{B}^n$ is another function that fulfills tcpo, then $\Phi \circ \Psi$ might not fulfill tcpo.

Theorem 9.4 For $\Phi : \mathbf{B}^n \longrightarrow \mathbf{B}^n$, we have
(a) $\forall \mu \in \mathbf{B}^n, \forall \nu \in \mathbf{B}^n, \Phi^{\nu}_{\mu} \subset \Phi_{\mu}$;
(b) Φ fulfills tcpo$\Longleftrightarrow \forall \nu \in \mathbf{B}^n, \Phi^{\nu}$ fulfills tcpo.

Proof: (a) Let $\mu, \nu \in \mathbf{B}^n$ arbitrary and fixed. We infer

$$\begin{aligned}\Phi^{\nu}_{\mu} &= \{i | i \in \{1,\dots,n\}, \Phi^{\nu}_i(\mu) \neq \mu_i\}\\ &= \{i | i \in \{1,\dots,n\}, \nu_i = 1 \text{ and } \Phi_i(\mu) \neq \mu_i\}\\ &\subset \{i | i \in \{1,\dots,n\}, \Phi_i(\mu) \neq \mu_i\} = \Phi_{\mu}.\end{aligned}$$

(b) $\Longrightarrow$ We take $\mu, \nu \in \mathbf{B}^n$ arbitrary. From (a) we get $card(\Phi^{\nu}_{\mu}) \leq card(\Phi_{\mu})$ and as $card(\Phi_{\mu}) \in \{0,1\}$, we infer $card(\Phi^{\nu}_{\mu}) \in \{0,1\}$.
$\Longleftarrow$ The implication is obvious if we take $\nu = (1,\dots,1) \in \mathbf{B}^n$ □

9.4 The Sets of Predecessors and Successors

Theorem 9.5 Φ fulfills tcpo if and only if $\forall \mu \in \mathbf{B}^n$,

$$\mu^- = \{\mu\} \cup \Phi^{-1}(\mu), \tag{9.12}$$

and also if and only if $\forall \mu \in \mathbf{B}^n$,

$$\mu^+ = \{\mu\} \cup \{\Phi(\mu)\}. \tag{9.13}$$

Proof: Let $\mu \in \mathbf{B}^n$ arbitrary.
(9.12). If. We take $\nu \in \mu^-$, $\nu \neq \mu$ arbitrary also and, from (9.12), we get $\Phi(\nu) = \mu$. We suppose against all reason that tcpo does not hold, thus $p \geq 2$ and $i_1, \dots, i_p \in \{1,\dots,n\}$ exist such that $\nu = \mu \oplus \varepsilon^{i_1} \oplus \dots \oplus \varepsilon^{i_p}$ and, on the other hand, $\lambda \in \mathbf{B}^n$ exists also with (see Theorem 1.5 (b), page 9)

$$\Phi^{\lambda}(\nu) = \mu \oplus \overline{\lambda_{i_1}}\varepsilon^{i_1} \oplus \dots \oplus \overline{\lambda_{i_p}}\varepsilon^{i_p} = \mu,$$

where $\lambda_{i_1} = \dots = \lambda_{i_p} = 1$.

We define $\beta \in \mathbf{B}^n$ by $\forall i \in \{1, \ldots, n\}$,

$$\beta_i = \begin{cases} \lambda_i, & \text{if } i \neq i_1, \\ 0, & \text{if } i = i_1 \end{cases}$$

and we have $\Phi^\beta(\nu) = \mu \oplus \varepsilon^{i_1}$. The hypothesis $(\mu \oplus \varepsilon^{i_1})^- = \{\mu \oplus \varepsilon^{i_1}\} \cup \Phi^{-1}(\mu \oplus \varepsilon^{i_1})$, together with $\nu \in (\mu \oplus \varepsilon^{i_1})^-$ imply $\Phi(\nu) = \mu \oplus \varepsilon^{i_1}$, contradiction.

Only if. The inclusion $\{\mu\} \cup \Phi^{-1}(\mu) \subset \mu^-$ holds, see Theorem 1.8, page 16. We prove $\mu^- \subset \{\mu\} \cup \Phi^{-1}(\mu)$.

We take an arbitrary $\delta \in \mu^-$ and we have two possibilities.

Case $\delta = \mu$

The inclusion is obvious.

Case $\delta \neq \mu$

Some $\lambda \in \mathbf{B}^n$ exists with $\Phi^\lambda(\delta) = \mu$ and some $i \in \{1, \ldots, n\}$ exists also such that

$$\Phi(\delta) = \delta \oplus \varepsilon^i \tag{9.14}$$

(otherwise, $\Phi(\delta) = \delta$ and $\delta = \Phi^\lambda(\delta) = \mu$ are in contradiction with the fact that $\delta \neq \mu$). We infer from Theorem 1.5 (a), page 9 that

$$\Phi^\lambda(\delta) = \delta \oplus \lambda_i \varepsilon^i. \tag{9.15}$$

As $\delta \neq \mu$, Eq. (9.15) corresponds to $\lambda_i = 1$ wherefrom $\mu = \delta \oplus \varepsilon^i$. We get from (9.14) that $\delta \in \Phi^{-1}(\delta \oplus \varepsilon^i) = \Phi^{-1}(\mu)$.

(9.13). If. We suppose against all reason that tcpo is false, i.e. $i_1, \ldots, i_p \in \{1, \ldots, n\}$ distinct exist, $p \geq 2$ with $\Phi(\mu) = \mu \oplus \varepsilon^{i_1} \oplus \ldots \oplus \varepsilon^{i_p}$. Then $\mu^+ = [\mu, \Phi(\mu)]$ gives $card(\mu^+) \geq 4$, contradiction with (9.13) that implies $card(\mu^+) \in \{1, 2\}$.

Only if. The possibility $\Phi(\mu) = \mu$ implies

$$\mu^+ = [\mu, \Phi(\mu)] = \{\mu\} = \{\mu\} \cup \{\Phi(\mu)\}$$

and the possibility $\exists i \in \{1, \ldots, n\}, \Phi(\mu) = \mu \oplus \varepsilon^i$ implies

$$\begin{aligned} \mu^+ &= [\mu, \Phi(\mu)] = [\mu, \mu \oplus \varepsilon^i] = \{\mu, \mu \oplus \varepsilon^i\} = \{\mu\} \cup \{\mu \oplus \varepsilon^i\} \\ &= \{\mu\} \cup \{\Phi(\mu)\} \end{aligned}$$

□

Theorem 9.6 Φ fulfills tcpo if and only if $\forall \mu \in \mathbf{B}^n$, one of

$$\mu^- = \{\mu\}, \tag{9.16}$$

$$\exists i \in \{1, \ldots, n\}, \mu^- = \{\mu, \mu \oplus \varepsilon^i\}, \tag{9.17}$$

$$\exists i \in \{1, \ldots, n\}, \exists j \in \{1, \ldots, n\}, \mu^- = \{\mu, \mu \oplus \varepsilon^i, \mu \oplus \varepsilon^j\}, \tag{9.18}$$

$$\vdots$$

$$\exists i_1 \in \{1, \ldots, n\}, \ldots, \exists i_n \in \{1, \ldots, n\}, \mu^- = \{\mu, \mu \oplus \varepsilon^{i_1}, \ldots, \mu \oplus \varepsilon^{i_n}\} \tag{9.19}$$

holds.

Proof: If. We suppose against all reason that this is not true, namely that $\mu \in \mathbf{B}^n, p \geq 2$ and $i_1, \dots, i_p \in \{1, \dots, n\}$ exist such that $\Phi(\mu \oplus \varepsilon^{i_1} \oplus \dots \oplus \varepsilon^{i_p}) = \mu$. This means that $\mu \oplus \varepsilon^{i_1} \oplus \dots \oplus \varepsilon^{i_p} \in \mu^-$, and in this situation (9.16)–(9.19) are all false, contradiction.

Only if. The statements (9.16)–(9.19) follow from Theorem 9.1, page 97 and Theorem 9.5: we replace in (9.12) the possible values of $\Phi^{-1}(\mu)$ resulting from (9.4)–(9.11) □

Theorem 9.7 If Φ fulfills tcpo, then $\forall \mu \in \mathbf{B}^n$, we have

$$O^-(\mu) = \{\mu\} \cup \Phi^{-1}(\mu) \cup \Phi^{-1}(\Phi^{-1}(\mu)) \cup \dots , \tag{9.20}$$

$$O^+(\mu) = \{\mu\} \cup \{\Phi(\mu)\} \cup \{\Phi^{(2)}(\mu)\} \cup \dots \tag{9.21}$$

Proof: (9.20). We start from $(1.29)_{\text{page 16}}$ and we apply (9.12). We compute:

$$\begin{aligned}\bigcup_{\lambda \in \mu^-} \lambda^- &= \bigcup_{\lambda \in \{\mu\} \cup \Phi^{-1}(\mu)} \{\lambda\} \cup \Phi^{-1}(\lambda) = \bigcup_{\lambda \in \{\mu\} \cup \Phi^{-1}(\mu)} \{\lambda\} \cup \bigcup_{\lambda \in \{\mu\} \cup \Phi^{-1}(\mu)} \Phi^{-1}(\lambda)\\ &= (\{\mu\} \cup \Phi^{-1}(\mu)) \cup (\Phi^{-1}(\mu) \cup \Phi^{-1}(\Phi^{-1}(\mu)))\\ &= \{\mu\} \cup \Phi^{-1}(\mu) \cup \Phi^{-1}(\Phi^{-1}(\mu))\end{aligned}$$

etc.

(9.21). We start from $(1.30)_{\text{page 16}}$ and we use (9.13). We can see that

$$\begin{aligned}\bigcup_{\lambda \in \mu^+} \lambda^+ &= \bigcup_{\lambda \in \{\mu\} \cup \{\Phi(\mu)\}} \{\lambda\} \cup \{\Phi(\lambda)\} = \bigcup_{\lambda \in \{\mu\} \cup \{\Phi(\mu)\}} \{\lambda\} \cup \bigcup_{\lambda \in \{\mu\} \cup \{\Phi(\mu)\}} \{\Phi(\lambda)\}\\ &= (\{\mu\} \cup \{\Phi(\mu)\}) \cup (\{\Phi(\mu)\} \cup \{\Phi^{(2)}(\mu)\})\\ &= \{\mu\} \cup \{\Phi(\mu)\} \cup \{\Phi^{(2)}(\mu)\}\end{aligned}$$

etc □

9.5 Source, Isolated Fixed Point, Transient Point, Sink

Theorem 9.8 We suppose that $\Phi : \mathbf{B}^n \longrightarrow \mathbf{B}^n$ fulfills tcpo. Then $\forall \mu \in \mathbf{B}^n$, the following possibilities exist:

(i) $\Phi^{-1}(\mu) = \emptyset$, when μ is a source, $\mu^- = \{\mu\}$ and $\exists i \in \{1, \dots, n\}, \mu^+ = \{\mu, \mu \oplus \varepsilon^i\}$;

(ii) $\Phi^{-1}(\mu) = \{\mu\}$, when μ is an isolated fixed point, $\mu^- = \{\mu\}$ and $\mu^+ = \{\mu\}$;

(iii) $\exists k \in \{1, \dots, n\}, \exists i_1 \in \{1, \dots, n\}, \dots, \exists i_k \in \{1, \dots, n\}, \Phi^{-1}(\mu) = \{\mu \oplus \varepsilon^{i_1}, \dots, \mu \oplus \varepsilon^{i_k}\}$, when μ is a transient point, $\mu^- = \{\mu, \mu \oplus \varepsilon^{i_1}, \dots, \mu \oplus \varepsilon^{i_k}\}$ and $\exists j \in \{1, \dots, n\}, \mu^+ = \{\mu, \mu \oplus \varepsilon^j\}$;

(iv) $\exists k \in \{1, \dots, n\}, \exists i_1 \in \{1, \dots, n\}, \dots, \exists i_k \in \{1, \dots, n\}, \Phi^{-1}(\mu) = \{\mu, \mu \oplus \varepsilon^{i_1}, \dots, \mu \oplus \varepsilon^{i_k}\}$, when μ is a sink, $\mu^- = \{\mu, \mu \oplus \varepsilon^{i_1}, \dots, \mu \oplus \varepsilon^{i_k}\}$ and $\mu^+ = \{\mu\}$.

Proof: We fix an arbitrary $\mu \in \mathbf{B}^n$ and we make use, when stating the possibilities (i)–(iv), of $(9.4)_{\text{page 98}}, \ldots, (9.11)_{\text{page 98}}$.

(i) $\Phi^{-1}(\mu) = \emptyset$;

We get from Theorem 9.5, page 101 that $\mu^- = \{\mu\}$.

In order to show the second statement, we know that two possibilities exist, from the fact that Φ fulfills tcpo: either $\Phi(\mu) = \mu$, which is not the case, since $\mu \in \Phi^{-1}(\mu)$ and then we get a contradiction with the supposition that $\Phi^{-1}(\mu) = \emptyset$, or $\exists i \in \{1, \ldots, n\}, \Phi(\mu) = \mu \oplus \varepsilon^i$, which proves that $\mu^+ = \{\mu, \mu \oplus \varepsilon^i\}$.

(ii) $\Phi^{-1}(\mu) = \{\mu\}$;

We infer from Theorem 9.5 that $\mu^- = \{\mu\}$.

The fact that $\mu^+ = [\mu, \Phi(\mu)] = [\mu, \mu] = \{\mu\}$ is obvious.

(iii) $\exists k \in \{1, \ldots, n\}, \exists i_1 \in \{1, \ldots, n\}, \ldots, \exists i_k \in \{1, \ldots, n\}, \Phi^{-1}(\mu) = \{\mu \oplus \varepsilon^{i_1}, \ldots, \mu \oplus \varepsilon^{i_k}\}$;

We have from Theorem 9.5 that $\mu^- = \{\mu, \mu \oplus \varepsilon^{i_1}, \ldots, \mu \oplus \varepsilon^{i_k}\}$.

The satisfaction of tcpo gives the next cases.

Case $\Phi(\mu) = \mu$

This is impossible, since it implies $\mu \in \Phi^{-1}(\mu) = \{\mu \oplus \varepsilon^{i_1}, \ldots, \mu \oplus \varepsilon^{i_k}\}$.

Case $\exists j \in \{1, \ldots, n\}, \Phi(\mu) = \mu \oplus \varepsilon^j$

We infer that $\mu^+ = [\mu, \Phi(\mu)] = \{\mu, \mu \oplus \varepsilon^j\}$ holds indeed.

(iv) $\exists k \in \{1, \ldots, n\}, \exists i_1 \in \{1, \ldots, n\}, \ldots, \exists i_k \in \{1, \ldots, n\}, \Phi^{-1}(\mu) = \{\mu, \mu \oplus \varepsilon^{i_1}, \ldots, \mu \oplus \varepsilon^{i_k}\}$;

Theorem 9.5 gives that $\mu^- = \{\mu, \mu \oplus \varepsilon^{i_1}, \ldots, \mu \oplus \varepsilon^{i_k}\}$.

As $\mu \in \Phi^{-1}(\mu)$, we get $\Phi(\mu) = \mu$ and $\mu^+ = [\mu, \Phi(\mu)] = \{\mu\}$ □

Corollary 9.1 If Φ is bijective and fulfills tcpo, then $\forall \mu \in \mathbf{B}^n$, one the following possibilities is true:

(j) $\Phi^{-1}(\mu) = \{\mu\}$, when μ is an isolated fixed point, $\mu^- = \{\mu\}$ and $\mu^+ = \{\mu\}$;

(jj) $\exists i \in \{1, \ldots, n\}, \Phi^{-1}(\mu) = \{\mu \oplus \varepsilon^i\}$, when μ is a transient point, $\mu^- = \{\mu, \mu \oplus \varepsilon^i\}$ and $\exists j \in \{1, \ldots, n\}, \mu^+ = \{\mu, \mu \oplus \varepsilon^j\}$.

Proof: This statement follows from Theorem 9.3, page 99 and Theorem 9.8 □

9.6 Isomorphisms vs tcpo

Theorem 9.9 We consider two functions $\Phi, \Psi : \mathbf{B}^n \longrightarrow \mathbf{B}^n$ and we suppose that $(h, h') \in Iso(\Phi, \Psi)$. If Φ fulfills tcpo and h is Lipschitz, i.e. $\forall \mu \in \mathbf{B}^n, \forall \lambda \in \mathbf{B}^n$,

$$d(h(\mu), h(\lambda)) \leq d(\mu, \lambda) \tag{9.22}$$

is true, where d is the Hamming distance (see Definition 2.3, page 28), then Ψ fulfills tcpo.

Proof: For any $\mu, \nu \in \mathbf{B}^n$, we can write

$$card(\Psi^{h'(\nu)}_{h(\mu)}) = d(h(\mu), \Psi^{h'(\nu)}(h(\mu))) = d(h(\mu), h(\Phi^{\nu}(\mu)))$$
$$\leq d(\mu, \Phi^{\nu}(\mu)) = card(\Phi^{\nu}_{\mu}) \in \{0, 1\}.$$

The statement $card(\Phi^{\nu}_{\mu}) \in \{0, 1\}$ follows from the fact that Φ fulfills tcpo and from Theorem 9.4 (b), page 101, implication $\Longrightarrow$. We have the existence of ν' such that $h'(\nu') = (1, \ldots, 1)$, for which $card(\Psi_{h(\mu)}) = card(\Psi^{h'(\nu')}_{h(\mu)}) \in \{0, 1\}$. When μ runs in $\mathbf{B}^n$, $h(\mu)$ runs in $\mathbf{B}^n$ also and therefore Ψ fulfills tcpo □

9.7 Antiisomorphisms vs tcpo

Theorem 9.10 Let the functions $\Phi, \Psi : \mathbf{B}^n \longrightarrow \mathbf{B}^n$ and $(h, h')^{\frown} \in Iso^{\frown}(\Phi, \Psi)$. We suppose that Φ fulfills tcpo, that h is Lipschitz and that the invariance property

$$\forall \nu \in \mathbf{B}^n, \Phi^{\nu}(\mathbf{B}^n) = \mathbf{B}^n \tag{9.23}$$

holds. Then Ψ satisfies tcpo and

$$\forall \nu \in \mathbf{B}^n, \Psi^{\nu}(\mathbf{B}^n) = \mathbf{B}^n. \tag{9.24}$$

Proof: For any $\mu, \nu \in \mathbf{B}^n$, we can write that

$$card(\Psi^{h'(\nu)}_{h(\Phi^{\nu}(\mu))}) = d(h(\Phi^{\nu}(\mu)), \Psi^{h'(\nu)}(h(\Phi^{\nu}(\mu)))) = d(h(\Phi^{\nu}(\mu)), h(\mu))$$
$$\leq d(\mu, \Phi^{\nu}(\mu)) = card(\Phi^{\nu}_{\mu}) \in \{0, 1\},$$

where $card(\Phi^{\nu}_{\mu}) \in \{0, 1\}$ results from the fact that Φ satisfies tcpo and from Theorem 9.4. We get the existence of ν' such that $h'(\nu') = (1, \ldots, 1)$, for which we have $card(\Psi_{h(\Phi^{\nu'}(\mu))}) = card(\Psi^{h'(\nu')}_{h(\Phi^{\nu'}(\mu))}) \in \{0, 1\}$. When μ runs in $\mathbf{B}^n$, $\Phi^{\nu'}(\mu)$ runs in $\mathbf{B}^n$ from (9.23), $h(\Phi^{\nu'}(\mu))$ runs itself in $\mathbf{B}^n$ since h is bijection and consequently Ψ satisfies tcpo. From the equation $h(\mu) = \Psi^{h'(\nu)}(h(\Phi^{\nu}(\mu)))$ with $\mu, \nu \in \mathbf{B}^n$ arbitrary and h, h', Φ^{ν} bijections, we infer the truth of (9.24) □

10

The Strong Technical Condition of Proper Operation

Strengthening tcpo is useful in discussing symmetry issues (see Chapter 14). The properties of the functions that satisfy tcpo are adapted to this context.

10.1 Definition

Theorem 10.1 For the function $\Phi : \mathbf{B}^n \longrightarrow \mathbf{B}^n$, the following statements are equivalent:

(a) Φ fulfills tcpo and $\forall \mu \in \mathbf{B}^n, \forall i \in \{1, \dots, n\}, \forall j \in \{1, \dots, n\}$,

$$(\Phi(\mu \oplus \varepsilon^i) = \mu \text{ and } \Phi(\mu \oplus \varepsilon^j) = \mu) \Longrightarrow (i = j); \tag{10.1}$$

(b) $\forall \mu \in \mathbf{B}^n$, the disjunction of the following four properties holds:

$$\Phi^{-1}(\mu) = \emptyset, \tag{10.2}$$

$$\Phi^{-1}(\mu) = \{\mu\}, \tag{10.3}$$

$$\exists i \in \{1, \dots, n\}, \Phi^{-1}(\mu) = \{\mu \oplus \varepsilon^i\}, \tag{10.4}$$

$$\exists i \in \{1, \dots, n\}, \Phi^{-1}(\mu) = \{\mu, \mu \oplus \varepsilon^i\}. \tag{10.5}$$

Proof: Let $\mu \in \mathbf{B}^n$ arbitrary and fixed.

(a) $\Rightarrow$ (b) As Φ fulfills tcpo, the disjunction of (10.2)–(10.5) or the disjunction of

$$\exists i \in \{1, \dots, n\}, \exists j \in \{1, \dots, n\}, \Phi^{-1}(\mu) = \{\mu \oplus \varepsilon^i, \mu \oplus \varepsilon^j\}, \tag{10.6}$$

$$\exists i \in \{1, \dots, n\}, \exists j \in \{1, \dots, n\}, \Phi^{-1}(\mu) = \{\mu, \mu \oplus \varepsilon^i, \mu \oplus \varepsilon^j\}, \tag{10.7}$$

$$\vdots$$

$$\exists i_1 \in \{1, \dots, n\}, \dots, \exists i_n \in \{1, \dots, n\}, \Phi^{-1}(\mu) = \{\mu \oplus \varepsilon^{i_1}, \dots, \mu \oplus \varepsilon^{i_n}\}, \tag{10.8}$$

$$\exists i_1 \in \{1, \dots, n\}, \dots, \exists i_n \in \{1, \dots, n\}, \Phi^{-1}(\mu) = \{\mu, \mu \oplus \varepsilon^{i_1}, \dots, \mu \oplus \varepsilon^{i_n}\} \tag{10.9}$$

Boolean Functions: Topics in Asynchronicity, First Edition. Serban E. Vlad.

holds. We suppose against all reason that (10.6) is true, thus $\exists i \in \{1, \ldots, n\}$, $\exists j \in \{1, \ldots, n\}$ such that $\Phi(\mu \oplus \varepsilon^i) = \mu, \Phi(\mu \oplus \varepsilon^j) = \mu$. The hypothesis states that $i = j$, contradiction. Similarly for (10.7).

$$\vdots$$

Let us suppose against all reason that (10.8) is true, thus $\exists i_1 \in \{1, \ldots, n\}, \ldots,$ $\exists i_n \in \{1, \ldots, n\}$ such that $\Phi(\mu \oplus \varepsilon^{i_1}) = \mu, \ldots, \Phi(\mu \oplus \varepsilon^{i_n}) = \mu$. The hypothesis implies that $i_1 = \ldots = i_n$, contradiction. Similarly for (10.9). Statement (b) holds.

(b) $\Rightarrow$ (a) The disjunction of (10.2)–(10.5) implies the satisfaction of (10.2) or …or (10.5) or (10.6) or …or (10.9) i.e. tcpo holds. Furthermore, we suppose against all reason that $\exists i \in \{1, \ldots, n\}, \exists j \in \{1, \ldots, n\}$ with

$$\Phi(\mu \oplus \varepsilon^i) = \mu \text{ and } \Phi(\mu \oplus \varepsilon^j) = \mu \text{ and } i \neq j.$$

This means the truth of $\mu \oplus \varepsilon^i, \mu \oplus \varepsilon^j \in \Phi^{-1}(\mu)$, contradiction with the hypothesis. Statement (a) takes place □

Definition 10.1 If one of the previous properties (a), (b) is true, we say that Φ fulfills the **strong technical condition of proper operation.**

Remark 10.1 The strong tcpo is useful in treating symmetry issues of the Boolean functions, as we shall see later.

Theorem 10.2 Φ fulfills the strong tcpo if and only if Φ^* fulfills the strong tcpo.

Proof: Only if. We take $\mu \in \mathbf{B}^n, i, j \in \{1, \ldots, n\}$ arbitrary and the hypothesis states the truth of (10.1), that is equivalent in succession with

$$\text{not } (\Phi(\mu \oplus \varepsilon^i) = \mu \text{ and } \Phi(\mu \oplus \varepsilon^j) = \mu) \text{ or } (i = j),$$
$$\Phi(\mu \oplus \varepsilon^i) \neq \mu \text{ or } \Phi(\mu \oplus \varepsilon^j) \neq \mu \text{ or } i = j,$$
$$\overline{\overline{\Phi(\overline{\overline{\mu \oplus \varepsilon^i}})}} \neq \overline{\overline{\mu}} \text{ or } \overline{\overline{\Phi(\overline{\overline{\mu \oplus \varepsilon^j}})}} \neq \mu \text{ or } i = j,$$
$$\overline{\Phi^*(\overline{\overline{\mu \oplus \varepsilon^i}})} \neq \overline{\overline{\mu}} \text{ or } \overline{\Phi^*(\overline{\overline{\mu \oplus \varepsilon^j}})} \neq \overline{\overline{\mu}} \text{ or } i = j,$$
$$\Phi^*(\overline{\mu \oplus \varepsilon^i}) \neq \overline{\mu} \text{ or } \Phi^*(\overline{\mu \oplus \varepsilon^j}) \neq \overline{\mu} \text{ or } i = j,$$
$$\Phi^*(\overline{\mu} \oplus \varepsilon^i) \neq \overline{\mu} \text{ or } \Phi^*(\overline{\mu} \oplus \varepsilon^j) \neq \overline{\mu} \text{ or } i = j,$$
$$\text{not } (\Phi^*(\overline{\mu} \oplus \varepsilon^i) = \overline{\mu} \text{ and } \Phi^*(\overline{\mu} \oplus \varepsilon^j) = \overline{\mu}) \text{ or } (i = j),$$
$$(\Phi^*(\overline{\mu} \oplus \varepsilon^i) = \overline{\mu} \text{ and } \Phi^*(\overline{\mu} \oplus \varepsilon^j) = \overline{\mu}) \Rightarrow (i = j).$$

When μ runs in $\mathbf{B}^n$, we have that $\overline{\mu}$ runs in $\mathbf{B}^n$ thus Φ^* fulfills the strong tcpo, if we take in consideration Theorem 9.2, page 98 also.

If. We run the previous proof conversely □

Theorem 10.3 If Φ is bijective, then the following statements are equivalent:
(a) $\forall \mu \in \mathbf{B}^n$, (10.3) or (10.4) is true;
(b) Φ fulfills tcpo;
(c) Φ fulfills the strong tcpo.

Proof: Theorem 9.3, page 99 shows that (a) $\Longleftrightarrow$ (b). On the other hand, the implications (a) $\Longrightarrow$ (c) $\Longrightarrow$ (b) are obvious □

10.2 Examples

Example 10.1 The identity $1_{\mathbf{B}^n} : \mathbf{B}^n \longrightarrow \mathbf{B}^n$ fulfills the strong tcpo.

Example 10.2 The function $\Phi : \mathbf{B}^2 \longrightarrow \mathbf{B}^2$ from Figure 10.1 fulfills also the strong tcpo:

$$\Phi^{-1}(0,1) = \Phi^{-1}(1,1) = \emptyset, \text{Eq. (10.2)},$$
$$\Phi^{-1}(0,0) = \{(0,0),(0,1)\}, \text{Eq. (10.5)},$$
$$\Phi^{-1}(1,0) = \{(1,0),(1,1)\}, \text{Eq. (10.5)}.$$

Figure 10.1 Φ fulfills the strong tcpo.

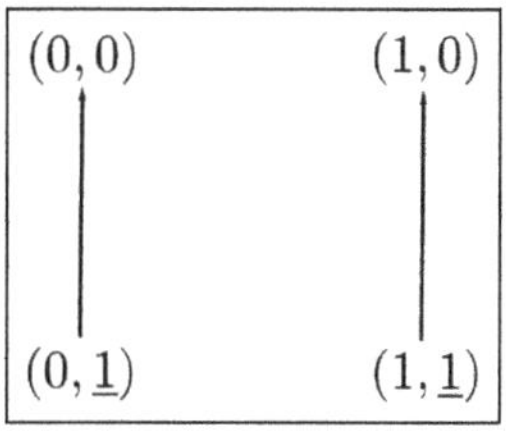

Example 10.3 The function $\Phi : \mathbf{B}^2 \longrightarrow \mathbf{B}^2$ from Figure 10.2 has the property that in all the points of $\mathbf{B}^2$ equation (10.4) holds, thus it fulfills the strong tcpo too.

Figure 10.2 Φ fulfills the strong tcpo.

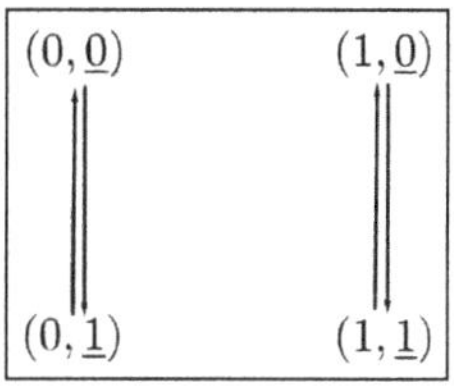

Example 10.4 We notice the fulfillment of the strong tcpo in the case of the function Φ from Figure 9.1, page 100:

$$\Phi^{-1}(0,0) = \emptyset, \text{Eq. (10.2)},$$
$$\Phi^{-1}(0,1) = \{(0,0)\}, \text{Eq. (10.4)},$$
$$\Phi^{-1}(1,1) = \{(0,1),(1,1)\}, \text{Eq. (10.5)},$$
$$\Phi^{-1}(1,0) = \{(1,0)\}, \text{Eq. (10.3)}.$$

Example 10.5 The function Φ from Figure 10.3 fulfills tcpo, but it does not fulfill the strong tcpo because $\Phi^{-1}(0,1) = \{(0,0),(1,1),(0,1)\}$, therefore all of (10.2)–(10.5) are false in (0, 1).

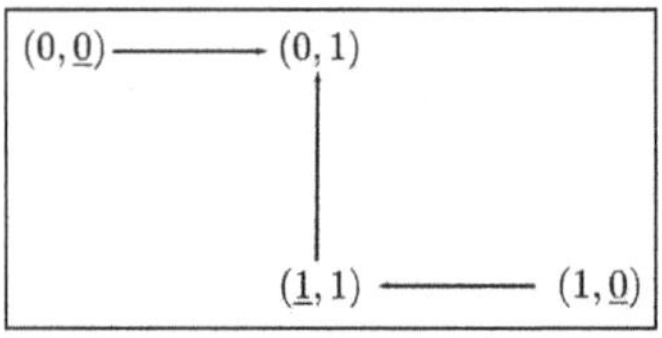

Figure 10.3 Φ fulfills tcpo but it does not fulfill the strong tcpo.

10.3 Iterates

Remark 10.2 If Φ fulfills the strong tcpo, then $\Phi \circ \Phi$ does not necessarily fulfill the strong tcpo, see Remark 9.2, page 101.

Theorem 10.4 Φ fulfills the strong tcpo$\Longleftrightarrow \forall \nu \in \mathbf{B}^n, \Phi^{\nu}$ fulfills the strong tcpo.

Proof: $\Rightarrow$ We take $\nu \in \mathbf{B}^n$ arbitrary and fixed. From Theorem 9.4 (b), page 101 we get that Φ^{ν} satisfies tcpo.

Let us suppose now that $\mu \in \mathbf{B}^n$ is arbitrary and $i \in \{1, \dots, n\}, j \in \{1, \dots, n\}$ exist such that $\Phi^{\nu}(\mu \oplus \varepsilon^i) = \mu, \Phi^{\nu}(\mu \oplus \varepsilon^j) = \mu$. We infer

$$\Phi_i^{\nu}(\mu \oplus \varepsilon^i) = \begin{cases} \mu_i \oplus 1, & \text{if} \quad \nu_i = 0, \\ \Phi_i(\mu \oplus \varepsilon^i), & \text{if} \quad \nu_i = 1 \end{cases} = \mu_i,$$

wherefrom $\nu_i = 1$ and $\Phi_i(\mu \oplus \varepsilon^i) = \mu_i$ are true. As Φ fulfills tcpo from the hypothesis, we obtain that $\Phi(\mu \oplus \varepsilon^i) = \mu$. The reasoning for getting $\Phi(\mu \oplus \varepsilon^j) = \mu$ is similar. We apply the hypothesis now that Φ fulfills the strong tcpo and the conclusion is $i = j$.

$\Longleftarrow$ This is obvious if we take $\nu = (1, \dots, 1) \in \mathbf{B}^n$ □

10.4 The Sets of Predecessors and Successors

Corollary 10.1 We suppose that Φ fulfills the strong tcpo. For any $\mu \in \mathbf{B}^n$,

(a) we have

$$\begin{aligned} \mu^- &= \{\mu\} \cup \Phi^{-1}(\mu), \\ \mu^+ &= \{\mu\} \cup \{\Phi(\mu)\}, \\ O^-(\mu) &= \{\mu\} \cup \Phi^{-1}(\mu) \cup \Phi^{-1}(\Phi^{-1}(\mu)) \cup \dots, \\ O^+(\mu) &= \{\mu\} \cup \{\Phi(\mu)\} \cup \{\Phi^{(2)}(\mu)\} \cup \dots; \end{aligned}$$

(b) either $\mu^- = \{\mu\}$, or $i \in \{1, \dots, n\}$ exists such that $\mu^- = \{\mu, \mu \oplus \varepsilon^i\}$. In a similar way, either $\mu^+ = \{\mu\}$, or $i \in \{1, \dots, n\}$ exists such that $\mu^+ = \{\mu, \mu \oplus \varepsilon^i\}$.

Proof: These assertions follow from Theorem 9.5, page 101, Theorem 9.7, page 103, and Theorem 10.1, page 107. □

10.5 Source, Isolated Fixed Point, Transient Point, Sink

Corollary 10.2 We suppose that Φ fulfills the strong tcpo. Then $\forall \mu \in \mathbf{B}^n$, one of the following statements is true:

(i) $\Phi^{-1}(\mu) = \emptyset$, μ is a source, $\mu^- = \{\mu\}$ and $\exists i \in \{1, \dots, n\}, \mu^+ = \{\mu, \mu \oplus \varepsilon^i\}$;

(ii) $\Phi^{-1}(\mu) = \{\mu\}$, μ is an isolated fixed point, $\mu^- = \{\mu\}$ and $\mu^+ = \{\mu\}$;

(iii) $\exists i \in \{1, \dots, n\}, \Phi^{-1}(\mu) = \{\mu \oplus \varepsilon^i\}$, μ is a transient point, $\mu^- = \{\mu, \mu \oplus \varepsilon^i\}$ and $\exists j \in \{1, \dots, n\}, \mu^+ = \{\mu, \mu \oplus \varepsilon^j\}$;

(iv) $\exists i \in \{1, \dots, n\}, \Phi^{-1}(\mu) = \{\mu, \mu \oplus \varepsilon^i\}$, μ is a sink, $\mu^- = \{\mu, \mu \oplus \varepsilon^i\}$ and $\mu^+ = \{\mu\}$.

Proof: This follows from Theorem 9.8, page 103 and (10.2)–(10.5). We use the fact that $\mu^- = \{\mu\} \cup \Phi^{-1}(\mu)$ □

Corollary 10.3 We suppose that Φ is bijective and it fulfills the strong tcpo. Then $\forall \mu \in \mathbf{B}^n$, one of (ii), (iii) from Corollary 10.2 holds.

Proof: This follows from Theorem 10.3, page 109. □

Example 10.6 In Figure 4.3, page 52 we have the example of a bijective function Φ that fulfills the strong tcpo: $(1, 0), (1, 1)$ are isolated fixed points and $(0, 0), (0, 1)$ are transient points.

10.6 Isomorphisms vs Strong tcpo

Theorem 10.5 Let the functions $\Phi, \Psi : \mathbf{B}^n \longrightarrow \mathbf{B}^n$ and we suppose that an isomorphism $(h, h') : \Phi \longrightarrow \Psi$ exists such that h is Lipschitz. If Φ satisfies the strong tcpo, then Ψ satisfies the strong tcpo.

Proof: Ψ satisfies tcpo from Theorem 9.9, page 104, thus Ψ^ν satisfies tcpo for any $\nu \in \mathbf{B}^n$, from Theorem 9.4, page 101. Let $\mu \in \mathbf{B}^n, \nu \in \mathbf{B}^n$ arbitrary and we suppose that $i \in \{1, \dots, n\}, j \in \{1, \dots, n\}$ exist such that

$$\Psi^{h'(\nu)}(\mu \oplus \varepsilon^i) = \mu \text{ and } \Psi^{h'(\nu)}(\mu \oplus \varepsilon^j) = \mu \tag{10.10}$$

holds. We denote with $\mu', \mu'', \mu''' \in \mathbf{B}^n$ the n-tuples defined by $h(\mu') = \mu \oplus \varepsilon^i, h(\mu'') = \mu \oplus \varepsilon^j, h(\mu''') = \mu$, in other words, from the commutativity of the diagram

$$\begin{array}{ccc} \mathbf{B}^n & \overset{\Phi^\nu}{\longrightarrow} & \mathbf{B}^n \\ h \downarrow & & \downarrow h \\ \mathbf{B}^n & \overset{\Psi^{h'(\nu)}}{\longrightarrow} & \mathbf{B}^n \end{array}$$

we infer that $\mu', \mu'' \in (\Phi^\nu)^{-1}(\mu''')$. We suppose against all reason that $\Psi^{h'(\nu)}$ does not satisfy the strong tcpo property, thus in (10.10) we have $i \neq j$. This implies $\mu \oplus \varepsilon^i \neq \mu \oplus \varepsilon^j$, thus $\mu' \neq \mu''$, since h is a bijection. As Φ^ν satisfies the strong tcpo from Theorem 10.4, the only possibility is given by the existence of $k \in \{1, \dots, n\}$ with $(\Phi^\nu)^{-1}(\mu''') = \{\mu''', \mu''' \oplus \varepsilon^k\}$, see Corollary 10.2. We can presume at this moment, without loosing the generality, that $\mu' = \mu'''$. We infer the contradiction

$$\mu = h(\mu''') = h(\mu') = \mu \oplus \varepsilon^i.$$

The conclusion is that $\Psi^{h'(\nu)}$ satisfies the strong tcpo. When ν runs in $\mathbf{B}^n$, as h' is bijective, $h'(\nu)$ runs in $\mathbf{B}^n$, therefore Ψ^ν satisfies the strong tcpo for any ν. We conclude from Theorem 10.4 that Ψ fulfills the strong tcpo □

10.7 Antiisomorphisms vs Strong tcpo

Theorem 10.6 We consider the functions $\Phi, \Psi : \mathbf{B}^n \longrightarrow \mathbf{B}^n$ and the antiisomorphism $(h, h')^{\frown} \in Iso^{\frown}(\Phi, \Psi)$. We suppose that Φ fulfills the strong tcpo, that h is Lipschitz and that the invariance property

$$\forall \nu \in \mathbf{B}^n, \Phi^\nu(\mathbf{B}^n) = \mathbf{B}^n \tag{10.11}$$

is also satisfied. Then Ψ satisfies the strong tcpo and

$$\forall \nu \in \mathbf{B}^n, \Psi^\nu(\mathbf{B}^n) = \mathbf{B}^n. \tag{10.12}$$

Proof: Theorem 9.10, page 105 shows that Ψ fulfills tcpo and (10.12), thus from Theorem 9.4, page 101 we get that Ψ^ν satisfies tcpo for any $\nu \in \mathbf{B}^n$. We take $\mu \in \mathbf{B}^n$, $\nu \in \mathbf{B}^n$ arbitrary for which

$$\Psi^{h'(\nu)}(\mu \oplus \varepsilon^i) = \mu \text{ and } \Psi^{h'(\nu)}(\mu \oplus \varepsilon^j) = \mu \tag{10.13}$$

holds, with $i \in \{1, \dots, n\}, j \in \{1, \dots, n\}$. From the commutativity of the diagram

$$\begin{array}{ccc} \mathbf{B}^n & \overset{\Phi^\nu}{\longrightarrow} & \mathbf{B}^n \\ h \downarrow & & \downarrow h \\ \mathbf{B}^n & \overset{\Psi^{h'(\nu)}}{\leftarrow} & \mathbf{B}^n \end{array}$$

in which h and Φ^{ν} are bijections, we get that $\Psi^{h'(\nu)}$ is a bijection, thus in (10.13) we have $\mu \oplus \varepsilon^{i} = \mu \oplus \varepsilon^{j}$, wherefrom $i = j$. It has resulted that $\Psi^{h'(\nu)}$ satisfies the strong tcpo. Similarly with the proof of Theorem 10.5, when ν runs in $\mathbf{B}^{n}$, the bijectivity of h' implies that $h'(\nu)$ runs in $\mathbf{B}^{n}$, therefore Ψ^{ν} satisfies the strong tcpo for any ν. In these circumstances, Theorem 10.4 states that Ψ fulfills the strong tcpo □

11

The Generalized Technical Condition of Proper Operation

If Φ fulfills tcpo, then the modeled circuit has a predictable behavior: $O^+(\mu) = \{\mu\} \cup \{\Phi(\mu)\} \cup \{\Phi^{(2)}(\mu)\} \cup \ldots$ Our purpose in this chapter is to generalize tcpo by allowing the existence of races, and still get predictability, under the form: $O^+(\mu) = [\mu, \Phi(\mu)] \cup [\Phi(\mu), \Phi^{(2)}(\mu)] \cup \ldots$

Φ fulfills the generalized tcpo$\Longleftrightarrow \forall \nu \in \mathbf{B}^n, \Phi^\nu$ fulfills the generalized tcpo.

If Φ fulfills the generalized tcpo, then $\forall \mu \in \mathbf{B}^n, \forall \nu \in \mu^-, [\nu, \mu] \subset \mu^-$ holds.

The form of $\Phi^{-1}(\mu)$ specific to the generalized tcpo characterizes the sources, the isolated fixed points, the transient points, and the sinks.

The isomorphisms and the antiisomorphisms bring, in certain circumstances, functions that fulfill the generalized tcpo, in functions that fulfill the generalized tcpo.

11.1 Definition

Theorem 11.1 For $\Phi : \mathbf{B}^n \longrightarrow \mathbf{B}^n$, the following statements (11.1)–(11.4) are equivalent:

$$\forall \mu \in \mathbf{B}^n, \forall \omega \in [\mu, \Phi(\mu)), \Phi(\mu) = \Phi(\omega), \tag{11.1}$$

$$\forall \mu \in \mathbf{B}^n, [\mu, \Phi(\mu)) \subset \Phi^{-1}(\Phi(\mu)), \tag{11.2}$$

$$\forall \nu \in \mathbf{B}^n, \forall \mu \in \Phi^{-1}(\nu), [\mu, \nu) \subset \Phi^{-1}(\nu), \tag{11.3}$$

$$\begin{aligned}&\forall \mu \in \mathbf{B}^n, \forall k \in \{2, \ldots, n\}, \forall i_1 \in \{1, \ldots, n\}, \ldots, \forall i_k \in \{1, \ldots, n\},\\ &\Phi(\mu) = \mu \oplus \varepsilon^{i_1} \oplus \ldots \oplus \varepsilon^{i_k}\\ &\Rightarrow \forall \lambda \in \mathbf{B}^k \setminus \{(1, \ldots, 1)\}, \Phi(\mu) = \Phi(\mu \oplus \lambda_1\varepsilon^{i_1} \oplus \ldots \oplus \lambda_k\varepsilon^{i_k})\end{aligned} \tag{11.4}$$

and any of them is equivalent with: $\forall \nu \in \mathbf{B}^n$, one of the following properties

$$\Phi^{-1}(\nu) = \emptyset, \tag{11.5}$$

$$\Phi^{-1}(\nu) = \{\nu\}, \tag{11.6}$$

$$\exists \lambda \in \mathbf{B}^n, \Phi^{-1}(\nu) = [\lambda, \nu), \tag{11.7}$$

Boolean Functions: Topics in Asynchronicity, First Edition. Serban E. Vlad.

$$\exists \lambda \in \mathbf{B}^n, \Phi^{-1}(\nu) = [\lambda, \nu], \tag{11.8}$$
$$\exists \lambda^1 \in \mathbf{B}^n, \exists \lambda^2 \in \mathbf{B}^n, \Phi^{-1}(\nu) = [\lambda^1, \nu) \cup [\lambda^2, \nu), \tag{11.9}$$
$$\exists \lambda^1 \in \mathbf{B}^n, \exists \lambda^2 \in \mathbf{B}^n, \Phi^{-1}(\nu) = [\lambda^1, \nu] \cup [\lambda^2, \nu], \tag{11.10}$$
$$\vdots$$
$$\exists k \geq 2, \exists \lambda^1 \in \mathbf{B}^n, \dots, \exists \lambda^k \in \mathbf{B}^n, \Phi^{-1}(\nu) = [\lambda^1, \nu) \cup \dots \cup [\lambda^k, \nu), \tag{11.11}$$
$$\exists k \geq 2, \exists \lambda^1 \in \mathbf{B}^n, \dots, \exists \lambda^k \in \mathbf{B}^n, \Phi^{-1}(\nu) = [\lambda^1, \nu] \cup \dots \cup [\lambda^k, \nu] \tag{11.12}$$

is true.

Proof: The scheme of the proof is:

$$(11.1) \Longrightarrow (11.2) \Longrightarrow (11.3) \Longrightarrow (11.4) \Longrightarrow (11.1),$$
$$(11.3) \Longrightarrow \forall \nu \in \mathbf{B}^n, ((11.5) \text{ or } (11.6) \text{ or} \dots \text{or } (11.12)) \Longrightarrow (11.3).$$

(11.1) $\Longrightarrow$ (11.2) We take $\mu \in \mathbf{B}^n$ arbitrary. If $\Phi(\mu) = \mu$, then (11.2) is trivially true, thus we suppose that $\Phi(\mu) \neq \mu$ and let $\omega \in [\mu, \Phi(\mu))$ arbitrary. As $\Phi(\mu) = \Phi(\omega)$, we obtain $\omega \in \Phi^{-1}(\Phi(\mu))$.

(11.2) $\Longrightarrow$ (11.3) We take an arbitrary $\nu \in \mathbf{B}^n$. If $\Phi^{-1}(\nu) = \emptyset$, then (11.3) is trivially true, thus we can suppose that $\Phi^{-1}(\nu) \neq \emptyset$ and we take an arbitrary $\mu \in \Phi^{-1}(\nu)$. In the situation when $\mu = \nu$, $[\mu, \nu) = \emptyset$ and the property (11.3) holds trivially again, therefore we can suppose that $\mu \neq \nu$ and let $\omega \in [\mu, \nu) = [\mu, \Phi(\mu))$ arbitrary. We get $\omega \in \Phi^{-1}(\Phi(\mu)) = \Phi^{-1}(\nu)$.

(11.3) $\Longrightarrow$ (11.4) Let $\mu \in \mathbf{B}^n$ arbitrary. We denote $\nu = \Phi(\mu)$ and we suppose that for $k \in \{2, \dots, n\}$, $i_1 \in \{1, \dots, n\}, \dots, i_k \in \{1, \dots, n\}$ we have $\Phi(\mu) = \mu \oplus \varepsilon^{i_1} \oplus \dots \oplus \varepsilon^{i_k}$. We fix an arbitrary $\lambda \in \mathbf{B}^k \setminus \{(1, \dots, 1)\}$. Then $\mu \oplus \lambda_1 \varepsilon^{i_1} \oplus \dots \oplus \lambda_k \varepsilon^{i_k} \in [\mu, \mu \oplus \varepsilon^{i_1} \oplus \dots \oplus \varepsilon^{i_k}) = [\mu, \Phi(\mu))$, thus

$$\nu = \Phi(\mu) \overset{(11.3)}{=} \Phi(\mu \oplus \lambda_1 \varepsilon^{i_1} \oplus \dots \oplus \lambda_k \varepsilon^{i_k}).$$

(11.4) $\Longrightarrow$ (11.1) We take an arbitrary $\mu \in \mathbf{B}^n$.

Case $\Phi(\mu) = \mu$

We have $[\mu, \Phi(\mu)) = \emptyset$ and the property

$$\forall \omega \in [\mu, \Phi(\mu)), \Phi(\mu) = \Phi(\omega) \tag{11.13}$$

is trivially true.

Case $\Phi(\mu) = \mu \oplus \varepsilon^i, i \in \{1, \dots, n\}$

In this case, $[\mu, \Phi(\mu)) = [\mu, \mu \oplus \varepsilon^i) = \{\mu\}$ and (11.13) is trivially true once again.

Case $\Phi(\mu) = \mu \oplus \varepsilon^{i_1} \oplus \dots \oplus \varepsilon^{i_k}, k \in \{2, \dots, n\}, i_1, \dots, i_k \in \{1, \dots, n\}$

Let $\omega \in [\mu, \Phi(\mu))$ arbitrary and fixed. We get the existence of $\lambda \in \mathbf{B}^k \setminus \{(1, \dots, 1)\}$ with the property $\omega = \mu \oplus \lambda_1 \varepsilon^{i_1} \oplus \dots \oplus \lambda_k \varepsilon^{i_k}$ and we can write

$$\Phi(\mu) \overset{(11.4)}{=} \Phi(\mu \oplus \lambda_1 \varepsilon^{i_1} \oplus \dots \oplus \lambda_k \varepsilon^{i_k}) = \Phi(\omega).$$

We show now that (11.3) implies the fact that $\forall \nu \in \mathbf{B}^n$, the disjunction of (11.5)–(11.12) holds and let ν arbitrary, fixed.

(a) Case $\Phi(\nu) \neq \nu$

If $\Phi^{-1}(\nu) = \emptyset$, then (11.5) is true and the implication holds, thus we can suppose that $\Phi^{-1}(\nu) \neq \emptyset$.

Let $\mu^1 \in \Phi^{-1}(\nu)$ arbitrary, thus $[\mu^1, \nu) \subset \Phi^{-1}(\nu)$. We infer from the hypothesis the existence of $p \geq 1$ and $\mu^2, \dots, \mu^p \in \mathbf{B}^n$ such that

$$[\mu^1, \nu) \subsetneq [\mu^2, \nu) \subsetneq \dots \subsetneq [\mu^p, \nu) \subset \Phi^{-1}(\nu),$$
$$\forall \mu \in \mathbf{B}^n, \text{not } ([\mu^p, \nu) \subsetneq [\mu, \nu) \subset \Phi^{-1}(\nu)).$$

In such conditions, we define $\lambda^1 = \mu^p$. If $\Phi^{-1}(\nu) = [\lambda^1, \nu)$, then the implication holds, thus we can suppose that $\Phi^{-1}(\nu) \neq [\lambda^1, \nu)$ and let $\omega^1 \in \Phi^{-1}(\nu) \setminus [\lambda^1, \nu)$ arbitrary. We get $[\omega^1, \nu) \subset \Phi^{-1}(\nu)$ and, moreover, we infer from the hypothesis the existence of $p' \geq 1$ and $\omega^2, \dots, \omega^{p'} \in \mathbf{B}^n$ such that

$$[\omega^1, \nu) \subsetneq [\omega^2, \nu) \subsetneq \dots \subsetneq [\omega^{p'}, \nu) \subset \Phi^{-1}(\nu),$$
$$\forall \mu \in \mathbf{B}^n, \text{not } ([\omega^{p'}, \nu) \subsetneq [\mu, \nu) \subset \Phi^{-1}(\nu)),$$

therefore we can define $\lambda^2 = \omega^{p'}$. If $\Phi^{-1}(\nu) = [\lambda^1, \nu) \cup [\lambda^2, \nu)$ then the implication holds, thus we can suppose that $\Phi^{-1}(\nu) \neq [\lambda^1, \nu) \cup [\lambda^2, \nu)$ and let $\delta^1 \in \Phi^{-1}(\nu) \setminus ([\lambda^1, \nu) \cup [\lambda^2, \nu))$ arbitrary. We get $[\delta^1, \nu) \subset \Phi^{-1}(\nu) \dots$

In finitely many steps, we get the existence of $\lambda^k \in \mathbf{B}^n$ such that $\Phi^{-1}(\nu) = [\lambda^1, \nu) \cup \dots \cup [\lambda^k, \nu)$ and the implication holds.

(b) Case $\Phi(\nu) = \nu$

If $\Phi^{-1}(\nu) = \{\nu\}$, then (11.6) is true and the implication holds, thus we can suppose that $\Phi^{-1}(\nu) \neq \{\nu\}$.

Let $\mu^1 \in \Phi^{-1}(\nu) \setminus \{\nu\}$ arbitrary, thus $[\mu^1, \nu] \subset \Phi^{-1}(\nu)$. The hypothesis shows the existence of $p \geq 1$ and $\mu^2, \dots, \mu^p \in \mathbf{B}^n$ such that

$$[\mu^1, \nu] \subsetneq [\mu^2, \nu] \subsetneq \dots \subsetneq [\mu^p, \nu] \subset \Phi^{-1}(\nu),$$
$$\forall \mu \in \mathbf{B}^n, \text{not } ([\mu^p, \nu] \subsetneq [\mu, \nu] \subset \Phi^{-1}(\nu))$$

and we define $\lambda^1 = \mu^p \dots$ The proof continues similarly with Case (a), until we get all of $\lambda^1 \in \mathbf{B}^n, \dots, \lambda^k \in \mathbf{B}^n$ such that $\Phi^{-1}(\nu) = [\lambda^1, \nu] \cup \dots \cup [\lambda^k, \nu]$. The implication is proved.

We show that $\forall \nu \in \mathbf{B}^n$, the disjunction of (11.5)–(11.12) implies (11.3). Let for this $\nu \in \mathbf{B}^n$ arbitrary, fixed. If (11.5) is true, then the implication

$$\forall \mu \in \emptyset, [\mu, \nu) \subset \emptyset$$

is trivially true.

We suppose that (11.6) is true, when the only choice of $\mu \in \Phi^{-1}(\nu)$ is $\mu = \nu$ and (11.3) is true under the form

$$\emptyset \subset \{\nu\}.$$

The rest of the possibilities is represented by the disjunction of (11.7)–(11.12), when we choose $\mu \in \Phi^{-1}(\nu)$ arbitrarily. In this case, $\lambda \in \mathbf{B}^n$ exists such that $\mu \in [\lambda, \nu)$ and

$$[\mu, \nu) \subset [\lambda, \nu) \subset \Phi^{-1}(\nu)$$

are true, thus (11.3) holds □

Definition 11.1 If one of the equivalent statements from Theorem 11.1 holds, we say that Φ fulfills the **generalized technical condition of proper operation.**

Remark 11.1 For any μ, a unique λ exists in (11.7), (11.8) such that $\Phi^{-1}(\mu) = [\lambda, \mu), \Phi^{-1}(\mu) = [\lambda, \mu]$ take place (see Theorem 2.2, page 23). The unique existence of distinct $\lambda^1, \lambda^2, \dots, \lambda^k$ is also true in (11.9)–(11.12), modulo their order.

Remark 11.2 For any μ, the generalized tcpo refers to the situation when μ and $\nu = \Phi(\mu)$ differ on $k \geq 2$ coordinates, $i_1, \dots, i_k$; then the value $\Phi(\mu)$ is asked to be equal with the value of Φ in any intermediate value $\omega = \mu \oplus \lambda_1 \varepsilon^{i_1} \oplus \dots \oplus \lambda_k \varepsilon^{i_k}, \lambda \neq (1, \dots, 1) \in \mathbf{B}^k$ that might result by the computation of $\leq k - 1$ unstable coordinates $\mu_i, i \in \{i_1, \dots, i_k\}$.

Remark 11.3 We see that tcpo is indeed a special case of the generalized tcpo. This happens since, if μ and $\Phi(\mu)$ differ on 0 or 1 coordinates, then the hypothesis of (11.4) is false and the generalized tcpo is fulfilled.

Remark 11.4 Statement (11.5) is a special case of (11.7), when $\lambda = \nu$ and $[\lambda, \nu) = \emptyset$; similarly, (11.6) is a special case of (11.8) when $\lambda = \nu$ and $[\lambda, \nu] = \{\nu\}$. Such remarks may continue, since (11.7) is a special case of (11.9) when $\lambda^1 = \lambda^2$, etc. We have written (11.5)–(11.12) under that form in order to state the generalized tcpo in a most intuitive manner.

Theorem 11.2 Φ fulfills the generalized tcpo if and only if Φ^* fulfills the generalized tcpo.

Proof: Only if. For any $\mu \in \mathbf{B}^n$, $\omega \in [\mu, \Phi(\mu))$, we notice first of all, see Remark 2.3, page 25, that

$$\begin{aligned}\overline{[\mu, \Phi(\mu))} &= \overline{[\mu, \Phi(\mu)] \setminus \{\Phi(\mu)\}} = \overline{[\mu, \Phi(\mu)]} \setminus \overline{\{\Phi(\mu)\}} \\ &= [\overline{\mu}, \overline{\Phi(\mu)}] \setminus \{\overline{\Phi(\mu)}\} = [\overline{\mu}, \overline{\Phi(\mu)}) = [\overline{\mu}, \Phi^*(\overline{\mu})).\end{aligned}$$

We take $\mu \in \mathbf{B}^n$, $\omega \in [\mu, \Phi(\mu))$ arbitrary, fixed and we have

$$\Phi^*(\overline{\omega}) = \overline{\Phi(\omega)} = \overline{\Phi(\mu)} = \Phi^*(\overline{\mu}).$$

Moreover, when μ runs in $\mathbf{B}^n$ and ω runs in $[\mu, \Phi(\mu))$, $\overline{\mu}$ runs in $\mathbf{B}^n$ and $\overline{\omega}$ runs in $[\overline{\mu}, \Phi^*(\overline{\mu}))$. Φ^* fulfills the generalized tcpo.

If. The inverse reasoning is clear now □

Theorem 11.3 If the function $\Phi : \mathbf{B}^n \longrightarrow \mathbf{B}^n$ is bijective, then the following statements are equivalent:

(a) $\forall \mu \in \mathbf{B}^n, (\Phi^{-1}(\mu) = \{\mu\}$ or $\exists i \in \{1, \dots, n\}, \Phi^{-1}(\mu) = \{\mu \oplus \varepsilon^i\})$;

(b) Φ satisfies tcpo;

(c) Φ satisfies the strong tcpo;

(d) Φ satisfies the generalized tcpo.

Proof: (a) $\Longleftrightarrow$ (b) $\Longleftrightarrow$ (c) coincides with Theorem 10.3, page 109.

(b) $\Longrightarrow$ (d) obvious (see Remark 11.3).

(d) $\Longrightarrow$ (b) We take $\mu \in \mathbf{B}^n$ arbitrary and fixed. If we would have, against all reason, $card([\mu, \Phi(\mu))) \geq 2$, then $\omega \in [\mu, \Phi(\mu))$ would exist, $\omega \neq \mu$, with $\Phi(\omega) = \Phi(\mu)$, contradiction with the bijectivity of Φ, thus $card([\mu, \Phi(\mu))) \in \{0, 1\}$. We have:

$$card([\mu, \Phi(\mu))) = card([\mu, \Phi(\mu)]) - 1 = 2^{d(\mu, \Phi(\mu))} - 1 = 2^{card(\Phi_\mu)} - 1.$$

From $2^{card(\Phi_\mu)} \in \{1, 2\}$, we infer that $card(\Phi_\mu) \in \{0, 1\}$, thus Φ satisfies tcpo □

11.2 Examples

Example 11.1 The identity $1_{\mathbf{B}^n} : \mathbf{B}^n \longrightarrow \mathbf{B}^n$ satisfies $\forall \mu \in \mathbf{B}^n$, $(1_{\mathbf{B}^n})^{-1}(\mu) = \{\mu\}$ therefore (11.6) is true.

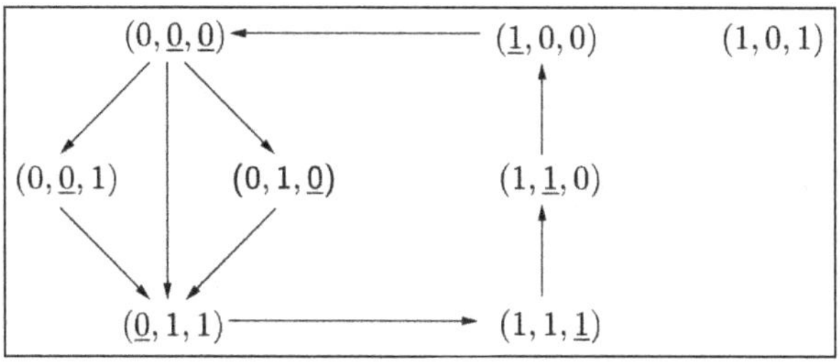

Figure 11.1 Φ fulfills the generalized tcpo.

Example 11.2 We give in Figure 11.1 an example of function Φ that fulfills the generalized tcpo. The most interesting computation here is $\Phi(0, 0, 0) = (0, 1, 1)$, which can take place in three different ways, as $\Phi_3(0, 0, 0)$ is computed first, and $(0, 0, 1)$ is an intermediate value; $\Phi_2(0, 0, 0)$ is computed first, and $(0, 1, 0)$ is an intermediate value; or $\Phi_2(0, 0, 0)$, $\Phi_3(0, 0, 0)$ are computed at the same time, and no intermediate values exist. All the other computations take place in similar conditions with tcpo.

Example 11.3 The function from Figure 11.2 fulfills the generalized tcpo.

Example 11.4 We have the example of the function from Figure 11.3 that fulfills also the generalized tcpo.

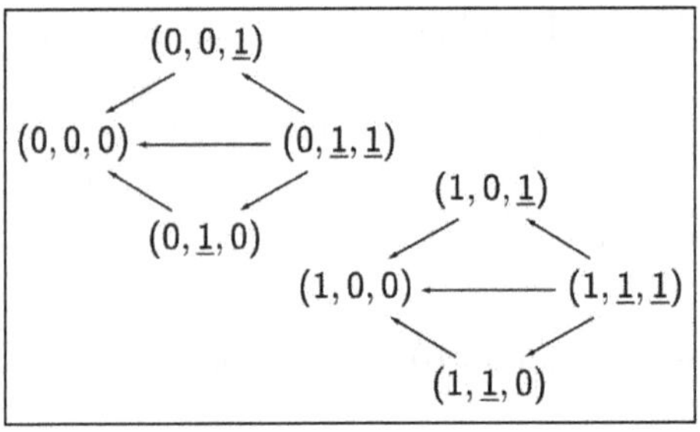

Figure 11.2 Φ fulfills the generalized tcpo.

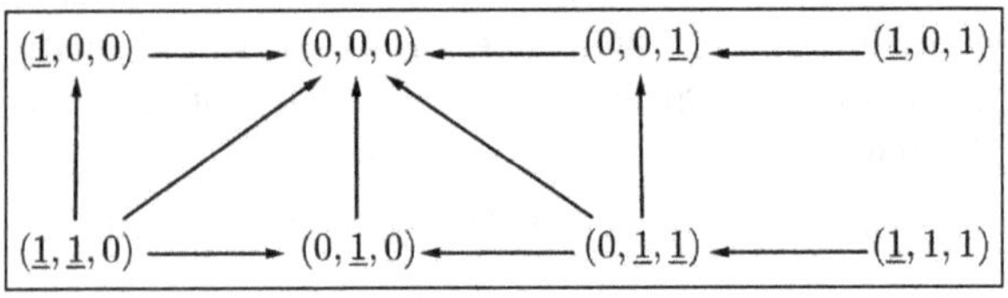

Figure 11.3 Φ fulfills the generalized tcpo.

Example 11.5 In Figure 1.4, page 13 we have the example of the function $\Phi : \mathbf{B}^2 \longrightarrow \mathbf{B}^2, \forall \mu \in \mathbf{B}^2$,

$$\Phi(\mu_1, \mu_2) = (\overline{\mu_1}, \overline{\mu_2})$$

that does not fulfill the generalized tcpo. This is seen from the counterexample: $\Phi(0, 0) = (1, 1)$, but $\Phi(0, 1) = (1, 0)$.

11.3 Iterates

Remark 11.5 We notice that if $\Phi : \mathbf{B}^n \longrightarrow \mathbf{B}^n$ fulfills the generalized tcpo, then $\Phi \circ \Phi$ might not fulfill the same property. For this, we denote with Φ the function from Figure 11.1. In (11.1) written for $\Phi \circ \Phi$ and $\mu = (1, 1, 1)$:

$$\forall \omega \in [(1, 1, 1), (1, 0, 0)), (\Phi \circ \Phi)(1, 1, 1) = (\Phi \circ \Phi)(\omega),$$

we take $\omega = (1, 0, 1)$. We get

$$(\Phi \circ \Phi)(1, 1, 1) = (1, 0, 0) \neq (1, 0, 1) = (\Phi \circ \Phi)(1, 0, 1).$$

More general, if $\Phi, \Psi : \mathbf{B}^n \longrightarrow \mathbf{B}^n$ fulfill the generalized tcpo, then $\Phi \circ \Psi$ might not fulfill the generalized tcpo.

Theorem 11.4 Φ fulfills the generalized tcpo if and only if for any $\lambda \in \mathbf{B}^n$, Φ^λ fulfills the generalized tcpo.

Proof: Only if. We fix $\lambda \in \mathbf{B}^n$, $\mu \in \mathbf{B}^n$ arbitrarily and we prove that

$$\forall \omega \in [\mu, \Phi(\mu)), \Phi(\mu) = \Phi(\omega) \tag{11.14}$$

implies

$$\forall \omega \in [\mu, \Phi^{\lambda}(\mu)), \Phi^{\lambda}(\mu) = \Phi^{\lambda}(\omega). \tag{11.15}$$

We suppose that $p \in \{1, \dots, n\}, i_1 \in \{1, \dots, n\}, \dots, i_p \in \{1, \dots, n\}$ exist such that

$$\Phi(\mu) = \mu \oplus \varepsilon^{i_1} \oplus \dots \oplus \varepsilon^{i_p}$$

and, from Theorem 1.5, page 9 we get

$$\Phi^{\lambda}(\mu) = \mu \oplus \lambda_{i_1}\varepsilon^{i_1} \oplus \dots \oplus \lambda_{i_p}\varepsilon^{i_p}.$$

In order that $\Phi^{\lambda}(\mu) \neq \mu$, for nontriviality, we have the existence of $j \in \{1, \dots, p\}$ with $\lambda_{i_j} = 1$. An element $\omega \in [\mu, \Phi^{\lambda}(\mu))$ fulfills

$$\omega = \mu \oplus \delta_{i_1}\lambda_{i_1}\varepsilon^{i_1} \oplus \dots \oplus \delta_{i_p}\lambda_{i_p}\varepsilon^{i_p}, \tag{11.16}$$

where $\delta \in \mathbf{B}^n$ and at least a $j \in \{1, \dots, p\}$ exists such that $\delta_{i_j} = 0, \lambda_{i_j} = 1$. As $\omega \in [\mu, \mu \oplus \varepsilon^{i_1} \oplus \dots \oplus \varepsilon^{i_p})$, we can apply (11.14) and we infer: $\forall k \in \{1, \dots, n\}$,

$$\Phi_k^{\lambda}(\omega) = \begin{cases} \omega_k, & \text{if } \lambda_k = 0, \\ \Phi_k(\omega), & \text{if } \lambda_k = 1 \end{cases} \overset{(11.14),(11.16)}{=} \begin{cases} \mu_k, & \text{if } \lambda_k = 0, \\ \Phi_k(\mu), & \text{if } \lambda_k = 1 \end{cases} = \Phi_k^{\lambda}(\mu).$$

If. This implication is obvious if we take $\lambda = (1, \dots, 1) \in \mathbf{B}^n$ □

11.4 The Sets of Predecessors and Successors

Remark 11.6 The properties $\forall \mu \in \mathbf{B}^n$,

$$\mu^- \supset \{\mu\} \cup \Phi^{-1}(\mu),$$
$$\mu^+ \supset \{\mu\} \cup \{\Phi(\mu)\},$$

see Theorem 1.8, page 16, which are to be compared with Theorem 9.5, page 101, do not hold as equality. In order to see this, we take a look for the first inclusion at Figure 11.1, page 119 where $(0,1,0)^- = \{(0,0,0), (0,1,0)\}$ and $\Phi^{-1}(0,1,0) = \emptyset$, thus $(0,1,0)^- \subset \{(0,1,0)\} \cup \Phi^{-1}(0,1,0)$ is false. In the same figure, we have $(0,0,0)^+ = \{(0,0,0), (0,0,1), (0,1,0), (0,1,1)\} \supset \{(0,0,0), (0,1,1)\} = \{(0,0,0)\} \cup \{\Phi(0,0,0)\}$.

Theorem 11.5 If Φ fulfills the generalized tcpo, then

$$\forall \mu \in \mathbf{B}^n, \forall \nu \in \mu^-, [\nu, \mu] \subset \mu^-.$$

Proof: Let $\mu \in \mathbf{B}^n, \nu \in \mu^-$ arbitrary and fixed. Some $\lambda \in \mathbf{B}^n$ exists with

$$\Phi^{\lambda}(\nu) = \mu. \tag{11.17}$$

If

$$\Phi(\nu) = \nu \tag{11.18}$$

then

$$\nu = \Phi(\nu) = \Phi^{\lambda}(\nu) \overset{(11.17)}{=} \mu$$

and the inclusion to be proved

$$[\nu, \mu] = [\nu, \nu] = \{\nu\} \subset \nu^{-}$$

is trivial, thus we can suppose from now the falsity of (11.18). In other words, $p, i_1, \dots, i_p \in \{1, \dots, n\}$ exist such that

$$\Phi(\nu) = \nu \oplus \varepsilon^{i_1} \oplus \dots \oplus \varepsilon^{i_p} \tag{11.19}$$

and we infer the truth of

$$\Phi^{\lambda}(\nu) \overset{(11.19)}{=} \nu \oplus \lambda_{i_1}\varepsilon^{i_1} \oplus \dots \oplus \lambda_{i_p}\varepsilon^{i_p} \overset{(11.17)}{=} \mu. \tag{11.20}$$

The satisfaction of the generalized tcpo means that

$$\forall \omega \in [\nu, \Phi(\nu)), \Phi(\nu) = \Phi(\omega) \tag{11.21}$$

and the inclusion to be proved is, from (11.20):

$$[\nu, \nu \oplus \lambda_{i_1}\varepsilon^{i_1} \oplus \dots \oplus \lambda_{i_p}\varepsilon^{i_p}] \subset \mu^{-}. \tag{11.22}$$

We take an arbitrary $\omega \in [\nu, \nu \oplus \lambda_{i_1}\varepsilon^{i_1} \oplus \dots \oplus \lambda_{i_p}\varepsilon^{i_p}]$, i.e. $\delta \in \mathbf{B}^n$ exists with

$$\omega = \nu \oplus \delta_{i_1}\lambda_{i_1}\varepsilon^{i_1} \oplus \dots \oplus \delta_{i_p}\lambda_{i_p}\varepsilon^{i_p} \tag{11.23}$$

and we must prove the existence of $\rho \in \mathbf{B}^n$ such that

$$\Phi^{\rho}(\omega) = \mu. \tag{11.24}$$

If $\omega = \nu \oplus \lambda_{i_1}\varepsilon^{i_1} \oplus \dots \oplus \lambda_{i_p}\varepsilon^{i_p} \overset{(11.20)}{=} \mu$, then equation (11.24) takes place for $\rho = (0, \dots, 0)$, thus we can suppose that $\omega \neq \mu$, in other words $\exists k \in \{1, \dots, p\}$ with $\delta_{i_k} = 0, \lambda_{i_k} = 1$. In these conditions $\omega \in [\nu, \Phi^{\lambda}(\nu)) \subset [\nu, \Phi(\nu))$ and we can apply (11.21). We have $\forall k \in \{1, \dots, n\}$,

$$\begin{aligned}
\Phi_k^{\rho}(\omega) &= \begin{cases} \omega_k, & \text{if } \rho_k = 0, \\ \Phi_k(\omega), & \text{if } \rho_k = 1 \end{cases} \\
&\overset{(11.21),(11.23)}{=} \begin{cases} \nu_k, & \text{if } k \in \{1, \dots, n\} \setminus \{i_1, \dots, i_p\}, \rho_k = 0, \\ \nu_k \oplus \delta_k\lambda_k, & \text{if } k \in \{i_1, \dots, i_p\}, \rho_k = 0, \\ \Phi_k(\nu), & \text{if } \rho_k = 1 \end{cases} \\
&\overset{(11.19)}{=} \begin{cases} \nu_k, & \text{if } k \in \{1, \dots, n\} \setminus \{i_1, \dots, i_p\}, \rho_k = 0, \\ \nu_k \oplus \delta_k\lambda_k, & \text{if } k \in \{i_1, \dots, i_p\}, \rho_k = 0, \\ \nu_k, & \text{if } k \in \{1, \dots, n\} \setminus \{i_1, \dots, i_p\}, \rho_k = 1, \\ \nu_k \oplus 1, & \text{if } k \in \{i_1, \dots, i_p\}, \rho_k = 1 \end{cases}
\end{aligned}$$

$$= \begin{cases} \nu_k, & \text{if } k \in \{1, \ldots, n\} \setminus \{i_1, \ldots, i_p\}, \\ \nu_k \oplus (\rho_k \cup \delta_k \lambda_k), & \text{if } k \in \{i_1, \ldots, i_p\} \end{cases} \overset{(11.24)}{=} \mu_k.$$

From (11.20), the last equality is true if we take $\rho = \lambda$. The inclusion (11.22) holds □

Theorem 11.6 We suppose that $\Phi : \mathbf{B}^n \longrightarrow \mathbf{B}^n$ fulfills the generalized tcpo and we take an arbitrary $\mu \in \mathbf{B}^n$. Then one of

$$\mu^- = \{\mu\}, \tag{11.25}$$

$$\exists \lambda \in \mathbf{B}^n, \mu^- = [\lambda, \mu], \tag{11.26}$$

$$\begin{aligned} &\exists k \in \{2, \ldots, 2^n\}, \exists \lambda^1 \in \mathbf{B}^n, \ldots, \exists \lambda^k \in \mathbf{B}^n, \\ &\mu^- = [\lambda^1, \mu] \cup \ldots \cup [\lambda^k, \mu] \end{aligned} \tag{11.27}$$

holds.

Proof: The proof is similar with the proof of (11.3) implies that $\forall \nu \in \mathbf{B}^n$, the disjunction of (11.5)–(11.12) holds from Theorem 11.1, page 115.

Let $\mu \in \mathbf{B}^n$ arbitrary, fixed. If μ is a source or an isolated fixed point, then (11.25) is true and the implication is proved, thus we can suppose for the rest of the proof that $\mu^- \neq \{\mu\}$.

We take $\mu^1 \in \mu^-$ and we get from Theorem 11.5 that $[\mu^1, \mu] \subset \mu^-$. We infer from the hypothesis the existence of $p \geq 1$ and $\mu^2, \ldots, \mu^p \in \mathbf{B}^n$ with

$$\begin{aligned} &[\mu^1, \mu] \subsetneq [\mu^2, \mu] \subsetneq \ldots \subsetneq [\mu^p, \mu] \subset \mu^-, \\ &\forall \nu \in \mathbf{B}^n, \text{not } ([\mu^p, \mu] \subsetneq [\nu, \mu] \subset \mu^-) \end{aligned}$$

and we define $\lambda^1 = \mu^p$. If $\mu^- = [\lambda^1, \mu]$, then (11.26) holds and the implication is true, thus we can suppose that $\mu^- \neq [\lambda^1, \mu]$ and let $\omega^1 \in \mu^- \setminus [\lambda^1, \mu]$ arbitrary. We obtain from Theorem 11.5 that $[\omega^1, \mu] \subset \mu^-$ and in addition the hypothesis states the existence of $p' \geq 1$ and $\omega^2, \ldots, \omega^{p'} \in \mathbf{B}^n$ with

$$\begin{aligned} &[\omega^1, \mu] \subsetneq [\omega^2, \mu] \subsetneq \ldots \subsetneq [\omega^{p'}, \mu] \subset \mu^-, \\ &\forall \nu \in \mathbf{B}^n, \text{not } ([\omega^{p'}, \mu] \subsetneq [\nu, \mu] \subset \mu^-) \end{aligned}$$

thus we can define $\lambda^2 = \omega^{p'}$. If $\mu^- = [\lambda^1, \mu] \cup [\lambda^2, \mu]$, (11.27) holds and the implication is true, therefore we can suppose that $\mu^- \neq [\lambda^1, \mu] \cup [\lambda^2, \mu]$ and let $\delta^1 \in \mu^- \setminus ([\lambda^1, \mu] \cup [\lambda^2, \mu])$ arbitrary. We have $[\delta^1, \mu] \subset \mu^- \ldots$

In finitely many steps, we obtain $\lambda^k \in \mathbf{B}^n$ such that (11.27) is true. The theorem is proved □

Remark 11.7 As we have previously noticed in a similar case, if $k = 1, \lambda^1 = \mu$, we get the special case $\mu^- = \{\mu\}$ and the above theorem could have been formulated in a more concise way.

Remark 11.8 The statement referring to the form of μ^+ when the generalized tcpo is fulfilled is trivial, since $\mu^+ = [\mu, \Phi(\mu)]$ is true irrespective of the fact that the generalized tcpo holds or not. Unlike the satisfaction of tcpo where $card(\mu^+) \in \{1, 2\}$, here we have $card(\mu^+) \in \{1, 2, \ldots, 2^n\}$, in particular $card(\mu^+) = 1$ if $\Phi(\mu) = \mu$ and $card(\mu^+) = 2^n$ if $\Phi(\mu) = \overline{\mu}$.

Theorem 11.7 If Φ fulfills the generalized tcpo, then $\forall \mu \in \mathbf{B}^n$ we have

$$O^-(\mu) \supset \{\mu\} \cup \Phi^{-1}(\mu) \cup \Phi^{-1}(\Phi^{-1}(\mu)) \cup \ldots, \tag{11.28}$$

$$O^+(\mu) = [\mu, \Phi(\mu)] \cup [\Phi(\mu), \Phi^{(2)}(\mu)] \cup \ldots \tag{11.29}$$

Proof: (11.28). Let $\nu \in \{\mu\} \cup \Phi^{-1}(\mu) \cup \Phi^{-1}(\Phi^{-1}(\mu)) \cup \ldots$ arbitrary. If $\nu = \mu$, then $\nu \in O^-(\mu)$, thus we can take $k \geq 1$ and $\nu \in \underbrace{\Phi^{-1}(\Phi^{-1}(\ldots(\Phi^{-1}}_{k}(\mu))\ldots))$, for which $\Phi^{(k)}(\nu) = \mu$, hence $\nu \in O^-(\mu)$.

(11.29). We prove

$$O^+(\mu) \subset [\mu, \Phi(\mu)] \cup [\Phi(\mu), \Phi^{(2)}(\mu)] \cup \ldots \tag{11.30}$$

and let $\nu \in O^+(\mu), \nu = (\Phi^{\lambda^p} \circ \ldots \circ \Phi^{\lambda^1})(\mu)$, where $\lambda^1, \ldots, \lambda^p \in \mathbf{B}^n$. If $\Phi^{\lambda^1}(\mu) = \Phi(\mu)$, we denote $i_1 = 1$. Otherwise $\Phi^{\lambda^1}(\mu) \in [\mu, \Phi(\mu))$, $\Phi(\Phi^{\lambda^1}(\mu)) = \Phi(\mu)$ and $\forall \nu \in \mathbf{B}^n, \Phi^{\nu}(\Phi^{\lambda^1}(\mu)) \in [\Phi^{\lambda^1}(\mu), \Phi(\mu)] \subset [\mu, \Phi(\mu)]$. If $(\Phi^{\lambda^2} \circ \Phi^{\lambda^1})(\mu) = \Phi(\mu)$, we denote $i_1 = 2$, otherwise $(\Phi^{\lambda^2} \circ \Phi^{\lambda^1})(\mu) \in [\mu, \Phi(\mu))$, $\Phi((\Phi^{\lambda^2} \circ \Phi^{\lambda^1})(\mu)) = \Phi(\mu)$ and $\forall \nu \in \mathbf{B}^n, \Phi^{\nu}((\Phi^{\lambda^2} \circ \Phi^{\lambda^1})(\mu)) \in [(\Phi^{\lambda^2} \circ \Phi^{\lambda^1})(\mu), \Phi(\mu)] \subset [\mu, \Phi(\mu)] \ldots$ We denote with $i_1 \geq 1$ the least number with $(\Phi^{\lambda^{i_1}} \circ \ldots \circ \Phi^{\lambda^1})(\mu) = \Phi(\mu)$ and we have

$$\forall k \in \{1, \ldots, i_1\}, (\Phi^{\lambda^k} \circ \ldots \circ \Phi^{\lambda^1})(\mu) \in [\mu, \Phi(\mu)].$$

If $(\Phi^{\lambda^{i_1+1}} \circ \ldots \circ \Phi^{\lambda^1})(\mu) = \Phi^{(2)}(\mu)$, we denote $i_2 = i_1 + 1$. Otherwise $(\Phi^{\lambda^{i_1+1}} \circ \ldots \circ \Phi^{\lambda^1})(\mu) \in [\Phi(\mu), \Phi^{(2)}(\mu))$, $\Phi((\Phi^{\lambda^{i_1+1}} \circ \ldots \circ \Phi^{\lambda^1})(\mu)) = \Phi^{(2)}(\mu)$ and we get $\forall \nu \in \mathbf{B}^n, \Phi^{\nu}((\Phi^{\lambda^{i_1+1}} \circ \ldots \circ \Phi^{\lambda^1})(\mu)) \in [(\Phi^{\lambda^{i_1+1}} \circ \ldots \circ \Phi^{\lambda^1})(\mu), \Phi^{(2)}(\mu)] \subset [\Phi(\mu), \Phi^{(2)}(\mu)] \ldots$ We denote with $i_2 \geq i_1 + 1$ the least number with $(\Phi^{\lambda^{i_2}} \circ \ldots \circ \Phi^{\lambda^1})(\mu) = \Phi^{(2)}(\mu)$ and we have

$$\forall k \in \{i_1, \ldots, i_2\}, (\Phi^{\lambda^k} \circ \ldots \circ \Phi^{\lambda^1})(\mu) \in [\Phi(\mu), \Phi^{(2)}(\mu)] \ldots$$

The reasoning is finite, and we finally get the existence of $i_q < p$ such that

$$\forall k \in \{i_q, \ldots, p\}, (\Phi^{\lambda^k} \circ \ldots \circ \Phi^{\lambda^1})(\mu) \in [\Phi^{(q-1)}(\mu), \Phi^{(q)}(\mu)].$$

The inclusion

$$O^+(\mu) \supset [\mu, \Phi(\mu)] \cup [\Phi(\mu), \Phi^{(2)}(\mu)] \cup \ldots$$

is obvious □

11.5 Source, Isolated Fixed Point, Transient Point, Sink

Theorem 11.8 Let $\Phi : \mathbf{B}^n \longrightarrow \mathbf{B}^n$. We suppose that the generalized tcpo holds and let $\mu \in \mathbf{B}^n$ be arbitrary, fixed. The following exclusive possibilities exist.

(i) If

$$\Phi^{-1}(\mu) = \emptyset,$$

then μ is either a source or a transient point;

(ii) if

$$\Phi^{-1}(\mu) = \{\mu\},$$

then μ is an isolated fixed point;

(iii) if

$$\exists p \in \{1, \dots, 2^n\}, \exists \lambda^1 \in \mathbf{B}^n, \dots, \exists \lambda^p \in \mathbf{B}^n,$$
$$\Phi^{-1}(\mu) = [\lambda^1, \mu) \cup \dots \cup [\lambda^p, \mu),$$

then μ is a transient point;

(iv) if

$$\exists p \in \{1, \dots, 2^n\}, \exists \lambda^1 \in \mathbf{B}^n, \dots, \exists \lambda^p \in \mathbf{B}^n,$$
$$\Phi^{-1}(\mu) = [\lambda^1, \mu] \cup \dots \cup [\lambda^p, \mu],$$

then μ is sink.

Proof: Case (i) In Figure 11.2, page 120 the generalized tcpo is satisfied. The point $\mu = (0, 1, 1)$ is a source with $\Phi^{-1}(0, 1, 1) = \emptyset$, $(0, 1, 1)^- = \{(0, 1, 1)\}$, $(0, 1, 1)^+ = \{(0, 1, 1), (0, 0, 1), (0, 1, 0), (0, 0, 0)\}$ and the point $\mu = (0, 0, 1)$ is transient, satisfying $\Phi^{-1}(0, 0, 1) = \emptyset$, $(0, 0, 1)^- = \{(0, 1, 1), (0, 0, 1)\}$, $(0, 0, 1)^+ = \{(0, 0, 1), (0, 0, 0)\}$. The first assertion of the theorem results from the fact that the isolated fixed points and the sinks μ satisfy $\Phi(\mu) = \mu$, thus $\mu \in \Phi^{-1}(\mu)$.

Case (ii) As $\Phi(\mu) = \mu$, we have $\mu^+ = \{\mu\}$ and we must still prove that $\mu^- = \{\mu\}$. We suppose against all reason that this is false, i.e. $\omega \neq \mu, \omega \in \mu^-$ exists, in other words we get the existence of $\nu \in \mathbf{B}^n$ such that $\Phi^\nu(\omega) = \mu$. We infer $\Phi(\omega) \neq \mu$ (otherwise $\Phi(\omega) = \mu$, resulting the contradiction $\omega \in \Phi^{-1}(\mu)$). The conclusion is $\mu \in (\omega, \Phi(\omega))$, but

$$\Phi(\omega) \overset{\text{gen tcpo}}{=} \Phi(\mu) = \mu$$

is a contradiction. This proves that $\mu^- = \{\mu\}$.

Case (iii) Theorem 1.8, page 16 shows that

$$\{\mu\} \neq \{\mu\} \cup [\lambda^1, \mu) \cup \dots \cup [\lambda^p, \mu) \subset \mu^-.$$

In addition, $\Phi(\mu) \neq \mu$ and $\mu^+ \neq \{\mu\}$ are clear.

Case (iv) Theorem 1.8 gives

$$\{\mu\} \neq \{\mu\} \cup [\lambda^1, \mu] \cup \ldots \cup [\lambda^p, \mu] \subset \mu^-.$$

Moreover, we infer $\Phi(\mu) = \mu$ and $\mu^+ = \{\mu\}$ □

Remark 11.9 To be compared (i)–(iv) of Theorem 11.8 with (i)–(iv) of Theorem 9.8, page 103 and to be noticed how the generalization works.

Remark 11.10 In Theorem 11.8, the situation $\Phi^{-1}(\mu) = [\mu, \Phi(\mu))$ is impossible. Indeed, there are two possibilities:

(a) Case $[\mu, \Phi(\mu)) = \emptyset$, when $\Phi(\mu) = \mu$. This implies that $\mu \in \Phi^{-1}(\mu)$, contradiction.

(b) Case $[\mu, \Phi(\mu)) \neq \emptyset$. As $\mu \in [\mu, \Phi(\mu))$, we obtain $\Phi(\mu) = \mu$, but this shows that $[\mu, \Phi(\mu)) = \emptyset$, contradiction.

Remark 11.11 If in Theorem 11.8 Φ is bijective, then it fulfills tcpo and one of the next statements is true (see Theorem 11.3, page 119) for any $\mu \in \mathbf{B}^n$:

(j) $\Phi^{-1}(\mu) = \{\mu\}$, when μ is an isolated fixed point;

(jj) $\exists i \in \{1, \ldots, n\}, \Phi^{-1}(\mu) = \{\mu \oplus \varepsilon^i\}$, when μ is a transient point.

11.6 Isomorphisms vs the Generalized tcpo

Theorem 11.9 We consider the functions $\Phi, \Psi : \mathbf{B}^n \longrightarrow \mathbf{B}^n$ and the isomorphism $(h, h') \in Iso(\Phi, \Psi)$. If Φ fulfills the generalized tcpo and h is compatible with the affine structure of $\mathbf{B}^n$, then Ψ fulfills the generalized tcpo □

Proof: For $\nu \in \mathbf{B}^n$ arbitrary, fixed the diagram

$$\begin{array}{ccc} \mathbf{B}^n & \stackrel{\Phi^{\nu}}{\longrightarrow} & \mathbf{B}^n \\ h \downarrow & & \downarrow h \\ \mathbf{B}^n & \stackrel{\Psi^{h'(\nu)}}{\longrightarrow} & \mathbf{B}^n \end{array}$$

is commutative. Let $\mu' \in \mathbf{B}^n$ arbitrary. If $\mu' = \Psi^{h'(\nu)}(\mu')$, then the property of generalized tcpo:

$$\forall \omega' \in [\mu', \Psi^{h'(\nu)}(\mu')), \Psi^{h'(\nu)}(\mu') = \Psi^{h'(\nu)}(\omega')$$

is trivially fulfilled, so that we can suppose from now that $\mu' \neq \Psi^{h'(\nu)}(\mu')$ and we take $\omega' \in [\mu', \Psi^{h'(\nu)}(\mu'))$ arbitrary itself. We define $\mu = h^{-1}(\mu')$ and

$\omega = h^{-1}(\omega')$. The fact that $h \in Af(\mathbf{B}^n)$ implies $h^{-1} \in Af(\mathbf{B}^n)$, see Theorem 2.7, page 28, thus

$$h^{-1}([\mu', \Psi^{h'(\nu)}(\mu')]) = [h^{-1}(\mu'), h^{-1}(\Psi^{h'(\nu)}(\mu'))] = [\mu, h^{-1}(\Psi^{h'(\nu)}(h(\mu)))]$$
$$= [\mu, h^{-1}(h(\Phi^{\nu}(\mu)))] = [\mu, \Phi^{\nu}(\mu)],$$

in particular $\omega \in [\mu, \Phi^{\nu}(\mu))$. But Φ^{ν} fulfills the generalized tcpo

$$\Phi^{\nu}(\mu) = \Phi^{\nu}(\omega)$$

from Theorem 11.4, page 120 and we infer

$$\Psi^{h'(\nu)}(\mu') = \Psi^{h'(\nu)}(h(\mu)) = h(\Phi^{\nu}(\mu)) = h(\Phi^{\nu}(\omega))$$
$$= \Psi^{h'(\nu)}(h(\omega)) = \Psi^{h'(\nu)}(\omega').$$

The previous property holds for any ν and any μ', with h' bijective, thus Ψ^{ν} fulfill all of them the generalized tcpo and we can apply Theorem 11.4 again in order to conclude that Ψ fulfills the generalized tcpo □

Example 11.6 The state portraits of two isomorphic functions $\Phi, \Psi : \mathbf{B}^3 \longrightarrow \mathbf{B}^3$ that fulfill the generalized tcpo have been drawn in Figures 11.4 and 11.5. The isomorphism $(\theta^{(0,0,1)}, 1_{\mathbf{B}^3}) \in Iso(\Phi, \Psi)$ satisfies the property that $\theta^{(0,0,1)}$ is compatible with the affine structure of $\mathbf{B}^3$, see Example 2.3, page 27.

Remark 11.12 We compare now Theorem 9.9, page 104 referring to tcpo with Theorem 11.9 referring to the generalized tcpo. At Theorem 9.9 the function h is bijective and Lipschitz, at Theorem 11.9 h is bijective and compatible with the affine structure of $\mathbf{B}^n$. Theorem 2.9, page 29 shows that in this last case h is Lipschitz with $\forall \mu \in \mathbf{B}^n, \forall \lambda \in \mathbf{B}^n, d(\mu, \lambda) = d(h(\mu), h(\lambda))$.

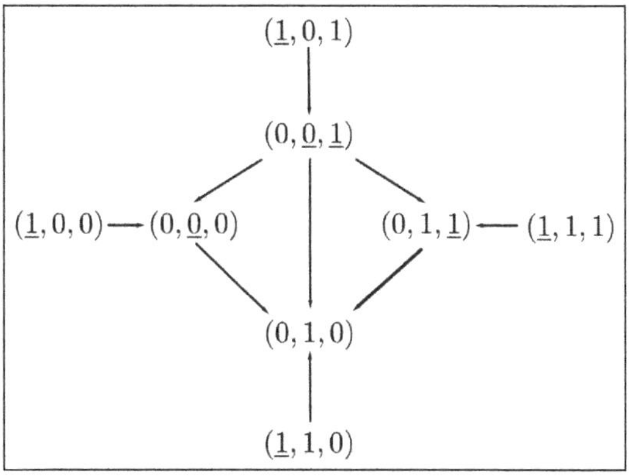

Figure 11.4 Function Φ that fulfills the generalized tcpo.

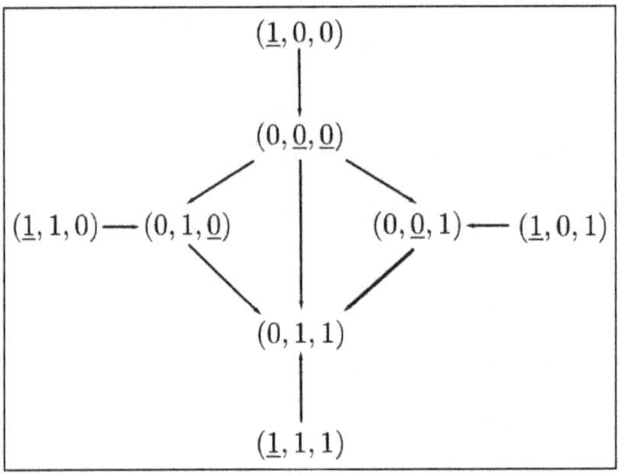

Figure 11.5 Function Ψ that fulfills the generalized tcpo.

11.7 Antiisomorphisms vs the Generalized tcpo

Theorem 11.10 Let $\Phi, \Psi : \mathbf{B}^n \longrightarrow \mathbf{B}^n$ and $(h, h')^{\frown} \in Iso^{\frown}(\Phi, \Psi)$. We suppose that Φ fulfills the generalized tcpo, that in addition $h \in Af(\mathbf{B}^n)$ and that the invariance property

$$\forall \nu \in \mathbf{B}^n, \Phi^{\nu}(\mathbf{B}^n) = \mathbf{B}^n \tag{11.31}$$

holds. Then Ψ satisfies

$$\forall \nu \in \mathbf{B}^n, \Psi^{\nu}(\mathbf{B}^n) = \mathbf{B}^n \tag{11.32}$$

and also the generalized tcpo.

Proof: We notice, first of all, that h is Lipschitz from Theorem 2.9, page 29. On the other hand (11.31) implies that Φ is bijective so that, taking into account the fact that it satisfies the generalized tcpo, we have from Theorem 11.3, page 119 that Φ fulfills tcpo. We can apply now Theorem 9.10, page 105, showing that Ψ fulfills tcpo and also the invariance (11.32), in particular Ψ is bijective. We conclude from Theorem 11.3 that Ψ fulfills the generalized tcpo □

Example 11.7 We get back to Example 4.4, page 47 where we have shown that the functions $\Phi(\mu_1, \mu_2) = (\mu_1, \overline{\mu_2})$ and $\Psi(\mu_1, \mu_2) = (\overline{\mu_1}, \mu_2)$ are antiisomorphic, by the existence of $(h, h')^{\frown} \in Iso^{\frown}(\Phi, \Psi)$, $h(\mu_1, \mu_2) = h'(\mu_1, \mu_2) = (\mu_2, \mu_1)$. Theorem 2.5, page 27 shows that $h \in Af(\mathbf{B}^n)$. In addition, Φ fulfills tcpo, the invariance (11.31) and, by Theorem 11.3, page 119, the generalized tcpo. We infer that Ψ satisfies the generalized tcpo in its special form represented by the invariance (11.32) and tcpo.

11.8 Other Properties

Theorem 11.11 Let $\mu \in \mathbf{B}^n$ and $p, i_1, \dots, i_p \in \{1, \dots, n\}$ such that

$$\Phi(\mu) = \mu \oplus \varepsilon^{i_1} \oplus \dots \oplus \varepsilon^{i_p}$$

and we suppose that Φ fulfills the generalized tcpo. Then $\forall \lambda \in \mathbf{B}^n, \forall \omega \in \mathbf{B}^n$, if $\omega_{i_1} \cdot \dots \cdot \omega_{i_p} = 0$, we have

$$\Phi^\lambda(\mu \oplus \omega_{i_1}\varepsilon^{i_1} \oplus \dots \oplus \omega_{i_p}\varepsilon^{i_p}) = \mu \oplus (\lambda_{i_1} \cup \omega_{i_1})\varepsilon^{i_1} \oplus \dots \oplus (\lambda_{i_p} \cup \omega_{i_p})\varepsilon^{i_p}.$$

Proof: We take $\lambda \in \mathbf{B}^n, \omega \in \mathbf{B}^n$ arbitrary such that $\omega_{i_1} \cdot \dots \cdot \omega_{i_p} = 0$. We denote $\nu = \Phi(\mu)$, wherefrom $\nu = \mu \oplus \varepsilon^{i_1} \oplus \dots \oplus \varepsilon^{i_p}$ and the generalized tcpo shows that $[\mu, \nu) \subset \Phi^{-1}(\nu)$. On the other hand $\mu \oplus \omega_{i_1}\varepsilon^{i_1} \oplus \dots \oplus \omega_{i_p}\varepsilon^{i_p} \in [\mu, \nu)$, thus we conclude

$$\Phi(\mu \oplus \omega_{i_1}\varepsilon^{i_1} \oplus \dots \oplus \omega_{i_p}\varepsilon^{i_p}) = \Phi(\mu). \tag{11.33}$$

We take $i \in \{1, \dots, n\}$ arbitrary and we infer

$$\Phi_i^\lambda(\mu \oplus \omega_{i_1}\varepsilon^{i_1} \oplus \dots \oplus \omega_{i_p}\varepsilon^{i_p})$$

$$= \begin{cases} \mu_i, i \in \{1, \dots, n\} \setminus \{i_1, \dots, i_p\}, \\ \Phi_i(\mu \oplus \omega_{i_1}\varepsilon^{i_1} \oplus \dots \oplus \omega_{i_p}\varepsilon^{i_p}), \lambda_i = 1, i \in \{i_1, \dots, i_p\}, \\ \mu_i \oplus \omega_i, \lambda_i = 0, i \in \{i_1, \dots, i_p\} \end{cases}$$

$$\overset{(11.33)}{=} \begin{cases} \mu_i, i \in \{1, \dots, n\} \setminus \{i_1, \dots, i_p\}, \\ \Phi_i(\mu), \lambda_i = 1, i \in \{i_1, \dots, i_p\}, \\ \mu_i, (\omega_i = 0 \text{ and } \lambda_i = 0), i \in \{i_1, \dots, i_p\}, \\ \mu_i \oplus 1, (\omega_i = 1 \text{ and } \lambda_i = 0), i \in \{i_1, \dots, i_p\} \end{cases}$$

$$= \begin{cases} \mu_i, i \in \{1, \dots, n\} \setminus \{i_1, \dots, i_p\}, \\ \mu_i \oplus 1, (\lambda_i = 1 \text{ or } (\omega_i = 1 \text{ and } \lambda_i = 0)), i \in \{i_1, \dots, i_p\}, \\ \mu_i, (\omega_i = 0 \text{ and } \lambda_i = 0), i \in \{i_1, \dots, i_p\} \end{cases}$$

$$= \begin{cases} \mu_i, i \in \{1, \dots, n\} \setminus \{i_1, \dots, i_p\}, \\ \mu_i \oplus (\lambda_i \cup \omega_i), i \in \{i_1, \dots, i_p\}. \end{cases}$$

The theorem is proved □

Theorem 11.12 If Φ fulfills the generalized tcpo, $\mu \in \mathbf{B}^n$ and $\omega \in [\mu, \Phi(\mu))$, then $\Phi_\omega \subset \Phi_\mu$.

Proof: From $\mu \oplus \underset{i \in \Phi_\mu}{\Xi} \varepsilon^i = \Phi(\mu) = \Phi(\omega) = \omega \oplus \underset{i \in \Phi_\omega}{\Xi} \varepsilon^i$ and from the existence of $A \subset \Phi_\mu$ with $\omega = \mu \oplus \underset{i \in A}{\Xi} \varepsilon^i$, we get $\underset{i \in A}{\Xi} \varepsilon^i = \underset{i \in \Phi_\mu}{\Xi} \varepsilon^i \oplus \underset{i \in \Phi_\omega}{\Xi} \varepsilon^i = \underset{i \in \Phi_\mu \Delta \Phi_\omega}{\Xi} \varepsilon^i$; $A = \Phi_\mu \Delta \Phi_\omega = (\Phi_\mu \setminus \Phi_\omega) \cup (\Phi_\omega \setminus \Phi_\mu) \subset \Phi_\mu$ implies $\Phi_\omega \setminus \Phi_\mu = \emptyset$ i.e. $\Phi_\omega \subset \Phi_\mu$ □

Theorem 11.13 If Φ satisfies the generalized tcpo, then $\forall \mu \in \mathbf{B}^n, \forall \mu' \in \mu^+$,

$$\Phi(\mu') = \mu' \Longrightarrow \mu' = \Phi(\mu). \tag{11.34}$$

Proof: We suppose against all reason that the property is false, thus $\mu \in \mathbf{B}^n$ and $\mu' \in [\mu, \Phi(\mu)]$ exist such that

$$\Phi(\mu') = \mu' \text{ and } \mu' \neq \Phi(\mu). \tag{11.35}$$

As $\mu' \in [\mu, \Phi(\mu))$, we apply the generalized tcpo and we infer

$$\mu' \overset{(11.35)}{=} \Phi(\mu') = \Phi(\mu),$$

contradiction with (11.35) □

Remark 11.13 We suppose that Φ fulfills the generalized tcpo. Then $\forall \mu \in \mathbf{B}^n, \forall \omega \in \mathbf{B}^n$,

$$\Phi(\mu) \neq \Phi(\omega) \Longrightarrow [\mu, \Phi(\mu)) \cap [\omega, \Phi(\omega)) = \emptyset.$$

If, against all reason, $\lambda \in [\mu, \Phi(\mu)) \cap [\omega, \Phi(\omega))$ would exist, then we would get the contradiction $\Phi(\mu) = \Phi(\lambda) = \Phi(\omega)$.

12

The Strong Generalized Technical Condition of Proper Operation

We strengthen the generalized tcpo to strong generalized tcpo for reasons related with symmetry that will become clear in Chapter 14, and we show how this new property degenerates if Φ is bijective.

If Φ fulfills the strong generalized tcpo, then $\Phi^{(k)}, k \in \mathbf{N}$ do not fulfill the same property, but $\Phi^{\lambda}, \lambda \in \mathbf{B}^n$ do.

We analyze the sources, the isolated fixed points, the transient points – which are two kinds: synchronous and asynchronous – and the sinks of Φ when it satisfies the strong generalized tcpo.

The isomorphisms and the antiisomorphisms preserve, in certain circumstances, the strong generalized tcpo.

12.1 Definition

Theorem 12.1 Let $\Phi : \mathbf{B}^n \longrightarrow \mathbf{B}^n$. Then $\forall \mu \in \mathbf{B}^n$,

$$\forall \omega \in [\mu, \Phi(\mu)), \Phi(\mu) = \Phi(\omega), \tag{12.1}$$

$$\begin{cases} \forall A \subset \{1, \ldots, n\}, \forall A' \subset \{1, \ldots, n\}, \\ (\Phi(\mu \oplus \underset{i \in A}{\Xi} \varepsilon^i) = \mu \text{ and } \Phi(\mu \oplus \underset{i \in A'}{\Xi} \varepsilon^i) = \mu) \Longrightarrow \Phi(\mu \oplus \underset{i \in A \cup A'}{\Xi} \varepsilon^i) = \mu \end{cases} \tag{12.2}$$

if and only if one of

$$\Phi^{-1}(\mu) = \emptyset, \tag{12.3}$$

$$\Phi^{-1}(\mu) = \{\mu\}, \tag{12.4}$$

$$\exists \lambda \in \mathbf{B}^n, \Phi^{-1}(\mu) = [\lambda, \mu), \tag{12.5}$$

$$\exists \lambda \in \mathbf{B}^n, \Phi^{-1}(\mu) = [\lambda, \mu] \tag{12.6}$$

holds.

Boolean Functions: Topics in Asynchronicity, First Edition. Serban E. Vlad.

Proof: We take an arbitrary $\mu \in \mathbf{B}^n$.

Only if. The statement (12.1) shows taking into account Theorem 11.1, page 115, that one of (12.3)–(12.6), or one of

$$\exists \lambda^1 \in \mathbf{B}^n, \exists \lambda^2 \in \mathbf{B}^n, \Phi^{-1}(\mu) = [\lambda^1, \mu) \cup [\lambda^2, \mu), \tag{12.7}$$

$$\exists \lambda^1 \in \mathbf{B}^n, \exists \lambda^2 \in \mathbf{B}^n, \Phi^{-1}(\mu) = [\lambda^1, \mu] \cup [\lambda^2, \mu], \tag{12.8}$$

$$\vdots$$

$$\exists k \geq 2, \exists \lambda^1 \in \mathbf{B}^n, \dots, \exists \lambda^k \in \mathbf{B}^n, \Phi^{-1}(\mu) = [\lambda^1, \mu) \cup \dots \cup [\lambda^k, \mu), \tag{12.9}$$

$$\exists k \geq 2, \exists \lambda^1 \in \mathbf{B}^n, \dots, \exists \lambda^k \in \mathbf{B}^n, \Phi^{-1}(\mu) = [\lambda^1, \mu] \cup \dots \cup [\lambda^k, \mu] \tag{12.10}$$

is true.

Case $\Phi(\mu) \neq \mu$

We suppose against all reason that the conclusion is false, i.e. $k \geq 2$ and $\lambda^1 \in \mathbf{B}^n, \dots, \lambda^k \in \mathbf{B}^n$ exist such that

$$\Phi^{-1}(\mu) = [\lambda^1, \mu) \cup \dots \cup [\lambda^k, \mu), \tag{12.11}$$

$$\forall i \in \{1, \dots, k\}, \forall j \in \{1, \dots, k\}, i \neq j \Longrightarrow [\lambda^i, \mu) \setminus [\lambda^j, \mu) \neq \emptyset. \tag{12.12}$$

We define the sets $A^i \subset \{1, \dots, n\}$ by

$$\lambda^i = \mu \oplus \underset{j \in A^i}{\Xi} \varepsilon^j, \tag{12.13}$$

$i \in \{1, \dots, k\}$. Since

$$\mu \overset{(12.11)}{=} \Phi(\lambda^i) \overset{(12.13)}{=} \Phi(\mu \oplus \underset{j \in A^i}{\Xi} \varepsilon^j),$$

we infer from (12.2) that

$$\Phi(\mu \oplus \underset{j \in A^1 \cup \dots \cup A^k}{\Xi} \varepsilon^j) = \mu.$$

We define $\lambda = \mu \oplus \underset{j \in A^1 \cup \dots \cup A^k}{\Xi} \varepsilon^j$, we have $\lambda \in \Phi^{-1}(\mu)$ and $(11.3)_{\text{page 115}}$ shows that $[\lambda, \mu) \subset \Phi^{-1}(\mu)$, i.e. $\lambda^1, \dots, \lambda^k \in [\lambda, \mu)$ and $[\lambda^1, \mu) \subset [\lambda, \mu), \dots, [\lambda^k, \mu) \subset [\lambda, \mu)$, representing a contradiction with (12.12). Statements (12.7)–(12.10) are all false.

Case $\Phi(\mu) = \mu$

The proof is similar, but the supposition against all reason refers to the existence of $k \geq 2$ and $\lambda^1 \in \mathbf{B}^n, \dots, \lambda^k \in \mathbf{B}^n$ such that

$$\Phi^{-1}(\mu) = [\lambda^1, \mu] \cup \dots \cup [\lambda^k, \mu],$$

$$\forall i \in \{1, \dots, k\}, \forall j \in \{1, \dots, k\}, i \neq j \Longrightarrow [\lambda^i, \mu] \setminus [\lambda^j, \mu] \neq \emptyset,$$

instead of (12.11), (12.12).

If. The disjunction of the statements (12.3)–(12.6) shows that the disjunction of (12.3)–(12.10) holds, thus (12.1) (the generalized tcpo) is true. Let $A', A'' \subset \{1, \dots, n\}$ with the property that $\lambda^1 = \mu \oplus \underset{i \in A'}{\Xi} \varepsilon^i$, $\lambda^2 = \mu \oplus \underset{i \in A''}{\Xi} \varepsilon^i$

satisfy $\Phi(\lambda^1) = \Phi(\lambda^2) = \mu$. The hypothesis shows the existence of $\lambda \in \mathbf{B}^n$ and $A \subset \{1, \dots, n\}$ such that $\lambda = \mu \oplus \underset{i\in A}{\Xi} \varepsilon^i$ and

$$\Phi^{-1}(\mu) = [\lambda, \mu) \text{ or } \Phi^{-1}(\mu) = [\lambda, \mu] \tag{12.14}$$

($\lambda = \mu$ is possible, thus (12.14) gives all the four possibilities (12.3)–(12.6)). We infer $\lambda^1, \lambda^2 \in [\lambda, \mu]$, wherefrom $A', A'' \subset A$. As $[\lambda, \mu] = \{\mu \oplus \underset{i\in H}{\Xi} \varepsilon^i | H \subset A\}$, we have in particular that $A' \cup A'' \subset A$, thus $\Phi(\mu \oplus \underset{i\in A'\cup A''}{\Xi} \varepsilon^i) = \mu$ □

Definition 12.1 We say that Φ fulfills the **strong generalized technical condition of proper operation** if $\forall \mu \in \mathbf{B}^n$,

$$\forall \lambda \in (\mu, \Phi(\mu)), \Phi^{-1}(\lambda) = \emptyset \tag{12.15}$$

and in addition one of the equivalent properties

(12.1) and (12.2)

(12.3) or (12.4) or (12.5) or (12.6)

is true.

Remark 12.1 Obviously, the existence of (12.1) between the previous requests shows that the strong generalized tcpo implies the generalized tcpo.

Remark 12.2 The conditions of proper operation (tcpo, strong tcpo, generalized tcpo, strong generalized tcpo) refer to computations of the function Φ that start in μ, include some possible intermediate values $\lambda \in (\mu, \Phi(\mu))$ and finalize in $\Phi(\mu)$.

Property (12.15) concerns the intermediate values λ and states that they cannot represent the finalization of a computation started in some point $\nu \in \Phi^{-1}(\lambda)$; thus, any intermediate value λ is an intermediate value only. We shall prove that these intermediate values are transient points, and we shall call them asynchronous transient points.

Theorem 12.2 Φ fulfills the strong generalized tcpo if and only if Φ^* fulfills the strong generalized tcpo.

Proof: Only if. For any $\mu \in \mathbf{B}^n, \lambda \in \mathbf{B}^n$, we have (see Remark 2.3, page 25):

$$\begin{aligned}\overline{(\mu, \Phi(\mu))} &= \overline{[\mu, \Phi(\mu)] \setminus \{\mu, \Phi(\mu)\}} = \overline{[\mu, \Phi(\mu)]} \setminus \overline{\{\mu, \Phi(\mu)\}} \\ &= [\overline{\mu}, \overline{\Phi(\mu)}] \setminus \{\overline{\mu}, \overline{\Phi(\mu)}\} = (\overline{\mu}, \overline{\Phi(\mu)}) = (\overline{\mu}, \Phi^*(\overline{\mu}))\end{aligned} \tag{12.16}$$

and similarly

$$\overline{[\lambda, \mu)} = \overline{[\lambda, \mu] \setminus \{\mu\}} = \overline{[\lambda, \mu]} \setminus \overline{\{\mu\}} = [\overline{\lambda}, \overline{\mu}] \setminus \{\overline{\mu}\} = [\overline{\lambda}, \overline{\mu}). \tag{12.17}$$

Let $\mu \in \mathbf{B}^n$ and $\lambda \in (\mu, \Phi(\mu))$ arbitrary. We have the following equivalent statements, in succession:

$$\Phi^{-1}(\lambda) = \emptyset,$$
$$\forall \nu \in \mathbf{B}^n, \Phi(\nu) \neq \lambda,$$
$$\forall \nu \in \mathbf{B}^n, \overline{\Phi(\overline{\overline{\nu}})} \neq \overline{\lambda},$$
$$\forall \nu \in \mathbf{B}^n, \Phi^*(\overline{\nu}) \neq \overline{\lambda},$$
$$(\Phi^*)^{-1}(\overline{\lambda}) = \emptyset,$$

where $\overline{\mu} \in \mathbf{B}^n, \overline{\lambda} \in \overline{(\mu\Phi(\mu))} = (\overline{\mu}, \Phi^*(\overline{\mu}))$.

Furthermore, the hypothesis states the truth of the disjunction of (12.3)–(12.6) that we prove to be equivalent with the disjunction of

$$(\Phi^*)^{-1}(\overline{\mu}) = \emptyset, \tag{12.18}$$
$$(\Phi^*)^{-1}(\overline{\mu}) = \{\overline{\mu}\}, \tag{12.19}$$
$$\exists \lambda \in \mathbf{B}^n, (\Phi^*)^{-1}(\overline{\mu}) = [\overline{\lambda}, \overline{\mu}), \tag{12.20}$$
$$\exists \lambda \in \mathbf{B}^n, (\Phi^*)^{-1}(\overline{\mu}) = [\overline{\lambda}, \overline{\mu}]. \tag{12.21}$$

For this, we choose to show the equivalence of (12.5) with (12.20). Indeed, we see that (12.5) is equivalent in succession with any of:

$$\exists \lambda \in \mathbf{B}^n, (\Phi^{-1}(\mu) \subset [\lambda, \mu) \text{ and } [\lambda, \mu) \subset \Phi^{-1}(\mu)),$$
$$\exists \lambda \in \mathbf{B}^n, ((\forall \nu \in \mathbf{B}^n, \Phi(\nu) = \mu \Longrightarrow \nu \in [\lambda, \mu)) \text{ and}$$
$$(\forall \nu, \nu \in [\lambda, \mu) \Longrightarrow \Phi(\nu) = \mu)),$$
$$\exists \lambda \in \mathbf{B}^n, ((\forall \nu \in \mathbf{B}^n, \Phi(\nu) \neq \mu \text{ or } \nu \in [\lambda, \mu)) \text{ and}$$
$$(\forall \nu, \nu \notin [\lambda, \mu) \text{ or } \Phi(\nu) = \mu)),$$
$$\exists \lambda \in \mathbf{B}^n, ((\forall \nu \in \mathbf{B}^n, \overline{\Phi(\overline{\overline{\nu}})} \neq \overline{\mu} \text{ or } \overline{\nu} \in \overline{[\lambda, \mu)}) \text{ and}$$
$$(\forall \nu, \overline{\nu} \notin \overline{[\lambda, \mu)} \text{ or } \overline{\Phi(\overline{\overline{\nu}})} = \overline{\mu})),$$
$$\exists \lambda \in \mathbf{B}^n, ((\forall \nu \in \mathbf{B}^n, \Phi^*(\overline{\nu}) \neq \overline{\mu} \text{ or } \overline{\nu} \in [\overline{\lambda}, \overline{\mu})) \text{ and}$$
$$(\forall \nu, \overline{\nu} \notin [\overline{\lambda}, \overline{\mu}) \text{ or } \Phi^*(\overline{\nu}) = \overline{\mu})),$$
$$\exists \lambda \in \mathbf{B}^n, ((\forall \nu \in \mathbf{B}^n, \Phi^*(\overline{\nu}) = \overline{\mu} \Longrightarrow \overline{\nu} \in [\overline{\lambda}, \overline{\mu})) \text{ and}$$
$$(\forall \nu, \overline{\nu} \in [\overline{\lambda}, \overline{\mu}) \Longrightarrow \Phi^*(\overline{\nu}) = \overline{\mu})),$$
$$\exists \lambda \in \mathbf{B}^n, ((\Phi^*)^{-1}(\overline{\mu}) \subset [\overline{\lambda}, \overline{\mu}) \text{ and}$$
$$[\overline{\lambda}, \overline{\mu}) \subset (\Phi^*)^{-1}(\overline{\mu})),$$

(12.20).

If. The steps of the proof are the same like previously, in the inverse order □

Theorem 12.3 If Φ fulfills the strong tcpo, then it fulfills also the strong generalized tcpo.

Proof: Let $\mu \in \mathbf{B}^n$ arbitrary. The disjunction of $(10.2)_{\text{page 107}}$–$(10.5)_{\text{page 107}}$ implies the disjunction of (12.3)–(12.6). On the other hand, as Φ fulfills tcpo, we get $(\mu, \Phi(\mu)) = \emptyset$, therefore (12.15) is trivially fulfilled □

Remark 12.3 The conclusion, as resulted by taking into account Theorem 12.3, is that the following implications hold:

$$\begin{array}{ccc} \text{tcpo} & \Longleftarrow & \text{strong tcpo} \\ \Downarrow & & \Downarrow \\ \text{generalized tcpo} & \Longleftarrow & \text{strong generalized tcpo} \end{array}$$

Theorem 12.4 If the function $\Phi : \mathbf{B}^n \longrightarrow \mathbf{B}^n$ is bijective, then the following statements are equivalent:

(a) $\forall \mu \in \mathbf{B}^n, (\Phi^{-1}(\mu) = \{\mu\}$ or $\exists i \in \{1, \ldots, n\}, \Phi^{-1}(\mu) = \{\mu \oplus \varepsilon^i\})$;
(b) Φ satisfies tcpo;
(c) Φ satisfies the strong tcpo;
(d) Φ satisfies the generalized tcpo;
(e) Φ fulfills the strong generalized tcpo.

Proof: (a) $\Longleftrightarrow$ (b) $\Longleftrightarrow$ (c) $\Longleftrightarrow$ (d) has been proved at Theorem 11.3, page 119.
(c) $\Longrightarrow$ (e) This is the statement of Theorem 12.3.
(e) $\Longrightarrow$ (d) obvious □

12.2 Examples

Example 12.1 The identity $1_{\mathbf{B}^n} : \mathbf{B}^n \longrightarrow \mathbf{B}^n$ fulfills the strong generalized tcpo.

Example 12.2 The function Φ from Figure 11.1, page 119 that fulfills the generalized tcpo fulfills also the strong generalized tcpo. For this, the satisfaction $\forall \mu \in \mathbf{B}^3$ of (12.15) is easily noticed:

$$(0,1,0) \in ((0,0,0),(0,1,1)) \text{ and } \Phi^{-1}(0,1,0) = \emptyset,$$
$$(0,0,1) \in ((0,0,0),(0,1,1)) \text{ and } \Phi^{-1}(0,0,1) = \emptyset$$

and the fulfillment of the disjunction of (12.3)–(12.6):

$$\Phi^{-1}(0,1,1) = [(0,0,0),(0,1,1)) = \{(0,0,0),(0,0,1),(0,1,0)\},$$
$$\Phi^{-1}(1,1,1) = [(0,1,1),(1,1,1)) = \{(0,1,1)\},$$
$$\Phi^{-1}(1,0,1) = \{(1,0,1)\},$$

etc., is clear too.

Example 12.3 The function from Figure 11.2, page 120 fulfills the strong generalized tcpo, similarly with the previous example.

Example 12.4 The function from Figure 11.3, page 120 fulfills the generalized tcpo, but it does not fulfill the strong generalized tcpo. For this it is enough to see that $(0, 0, 1) \in ((0, 1, 1), (0, 0, 0))$ and $\Phi^{-1}(0, 0, 1) = \{(1, 0, 1)\} \neq \emptyset$, i.e. (12.15) is false.

12.3 Iterates

Remark 12.4 If Φ fulfills the strong generalized tcpo, then $\Phi \circ \Phi$ might not fulfill the same property. It is sufficient in this respect to refer to Remark 11.5, page 120 and the example given there.

Theorem 12.5 Φ fulfills the strong generalized tcpo if and only if for any $\nu \in \mathbf{B}^n$, Φ^ν fulfills the strong generalized tcpo.

Proof: Only if. We fix arbitrarily $\mu \in \mathbf{B}^n$ and $\nu \in \mathbf{B}^n$. We prove that the conjunction of

$$\forall \lambda \in (\mu, \Phi(\mu)), \Phi^{-1}(\lambda) = \emptyset, \tag{12.22}$$

$$\forall \omega \in [\mu, \Phi(\mu)), \Phi(\mu) = \Phi(\omega), \tag{12.23}$$

$$\begin{cases} \forall A \subset \{1, \dots, n\}, \forall A' \subset \{1, \dots, n\}, \\ (\Phi(\mu \oplus \underset{i \in A}{\Xi} \varepsilon^i) = \mu \text{ and } \Phi(\mu \oplus \underset{i \in A'}{\Xi} \varepsilon^i) = \mu) \Longrightarrow \Phi(\mu \oplus \underset{i \in A \cup A'}{\Xi} \varepsilon^i) = \mu \end{cases} \tag{12.24}$$

implies the conjunction of

$$\forall \lambda \in (\mu, \Phi^\nu(\mu)), (\Phi^\nu)^{-1}(\lambda) = \emptyset, \tag{12.25}$$

$$\forall \omega \in [\mu, \Phi^\nu(\mu)), \Phi^\nu(\mu) = \Phi^\nu(\omega), \tag{12.26}$$

$$\begin{cases} \forall A \subset \{1, \dots, n\}, \forall A' \subset \{1, \dots, n\}, \\ (\Phi^\nu(\mu \oplus \underset{i \in A}{\Xi} \varepsilon^i) = \mu \text{ and } \Phi^\nu(\mu \oplus \underset{i \in A'}{\Xi} \varepsilon^i) = \mu) \\ \Longrightarrow \Phi^\nu(\mu \oplus \underset{i \in A \cup A'}{\Xi} \varepsilon^i) = \mu. \end{cases} \tag{12.27}$$

We notice first of all that the generalized tcpo (12.26) is true from Theorem 11.4, page 120.

We suppose against all reason that (12.25) is false, i.e. $\lambda \in (\mu, \Phi^\nu(\mu))$ and $\omega \in \mathbf{B}^n$ exist with $\Phi^\nu(\omega) = \lambda$. We have

$$\begin{aligned} (\mu, \Phi^\nu(\mu)) &= \{\mu \oplus \underset{i \in A}{\Xi} \varepsilon^i | \emptyset \subsetneq A \subsetneq \Phi^\nu_\mu\} \\ &= \{\mu \oplus \underset{i \in A}{\Xi} \varepsilon^i | \emptyset \subsetneq A \subsetneq \Phi_\mu \cap \{j | j \in \{1, \dots, n\}, \nu_j = 1\}\} \\ &\subset \{\mu \oplus \underset{i \in A}{\Xi} \varepsilon^i | \emptyset \subsetneq A \subsetneq \Phi_\mu\} = (\mu, \Phi(\mu)), \end{aligned}$$

thus $\lambda \in [\omega, \Phi(\omega)] \cap (\mu, \Phi(\mu))$. The possibility $\lambda = \Phi(\omega)$ is excluded, since it is in contradiction with (12.22), in other words $\lambda \in [\omega, \Phi(\omega)) \cap (\mu, \Phi(\mu))$ and we can apply the property of generalized tcpo of Φ, wherefrom $\Phi(\omega) = \Phi(\lambda) = \Phi(\mu)$.

As $\lambda \in (\mu, \Phi^{\nu}(\mu))$, we infer $\exists i \in \{1, \dots, n\}, \mu_i = \lambda_i \neq \Phi_i^{\nu}(\mu), \nu_i = 1$ and $\exists j \in \{1, \dots, n\}, \mu_j \neq \lambda_j = \Phi_j^{\nu}(\mu), \nu_j = 1$. The first of these remarks implies

$$\lambda_i = \Phi_i^{\nu}(\omega) = \Phi_i(\omega) = \Phi_i(\mu) = \Phi_i^{\nu}(\mu),$$

contradiction. Statement (12.25) is proved.

We prove (12.27) and we suppose that, for arbitrary $A \subset \{1, \dots, n\}, A' \subset \{1, \dots, n\}$,

$$\Phi^{\nu}(\mu \oplus \underset{i\in A}{\Xi}\, \varepsilon^i) = \mu \text{ and } \Phi^{\nu}(\mu \oplus \underset{i\in A'}{\Xi}\, \varepsilon^i) = \mu$$

is true. For any $j \in \{1, \dots, n\}$,

$$\Phi_j^{\nu}(\mu \oplus \underset{i\in A}{\Xi}\, \varepsilon^i) = \begin{cases} \mu_j, & \text{if } \nu_j = 0, j \notin A, \\ \mu_j \oplus 1, & \text{if } \nu_j = 0, j \in A, \\ \Phi_j(\mu \oplus \underset{i\in A}{\Xi}\, \varepsilon^i), & \text{if } \nu_j = 1 \end{cases} = \mu_j$$

implies

$$\begin{cases} \{i | i \in \{1, \dots, n\}, \nu_i = 0\} \cap A = \emptyset, \\ \forall j \in \{i | i \in \{1, \dots, n\}, \nu_i = 1\}, \Phi_j(\mu \oplus \underset{i\in A}{\Xi}\, \varepsilon^i) = \mu_j \end{cases}$$

and similarly

$$\begin{cases} \{i | i \in \{1, \dots, n\}, \nu_i = 0\} \cap A' = \emptyset, \\ \forall j \in \{i | i \in \{1, \dots, n\}, \nu_i = 1\}, \Phi_j(\mu \oplus \underset{i\in A'}{\Xi}\, \varepsilon^i) = \mu_j \end{cases}$$

is true too. We infer that

$$\begin{cases} \{i | i \in \{1, \dots, n\}, \nu_i = 0\} \cap (A \cup A') = \emptyset, \\ \forall j \in \{i | i \in \{1, \dots, n\}, \nu_i = 1\}, \Phi_j(\mu \oplus \underset{i\in A\cup A'}{\Xi}\, \varepsilon^i) = \mu_j \end{cases}$$

is true, and we have used (12.24). This implies

$$\Phi^{\nu}(\mu \oplus \underset{i\in A\cup A'}{\Xi}\, \varepsilon^i) = \mu.$$

The statement (12.27) is proved.

If. This implication is obvious, by taking $\nu = (1, \dots, 1) \in \mathbf{B}^n$ □

12.4 Source, Isolated Fixed Point, Transient Point, Sink

Theorem 12.6 If Φ fulfills the strong generalized tcpo then, for any $\mu \in \mathbf{B}^n$, the following exclusive possibilities (i)–(iv) exist:

(i) $\Phi^{-1}(\mu) = \emptyset, \Phi(\mu) \neq \mu, \mu^+ \neq \{\mu\}$ and either
(i.1) $\mu^- = \{\mu\}$, or
(i.2) $\exists \lambda \in \mathbf{B}^n$ such that

$$\Phi^{-1}(\Phi(\lambda)) = [\lambda, \Phi(\lambda)) \text{ or } \Phi^{-1}(\Phi(\lambda)) = [\lambda, \Phi(\lambda)], \tag{12.28}$$

$\mu \in (\lambda, \Phi(\lambda))$ and we have:

$$\{\mu\} \neq \mu^- = [\lambda, \mu] = \{\lambda \oplus \underset{i \in A}{\Xi} \varepsilon^i | A \subset (\lambda \boxplus \mu)\},$$

$$\mu^+ = [\mu, \Phi(\lambda)] = \{\lambda \oplus \underset{i \in A}{\Xi} \varepsilon^i | (\lambda \boxplus \mu) \subset A \subset \Phi_\lambda\};$$

(ii) $\Phi^{-1}(\mu) = \{\mu\}, \mu^- = \{\mu\}, \Phi(\mu) = \mu, \mu^+ = \{\mu\}$;
(iii) $\exists \lambda \in \mathbf{B}^n, \Phi^{-1}(\mu) = [\lambda, \mu)$, where $\lambda \neq \mu, \{\mu\} \neq \mu^- = [\lambda, \mu], \Phi(\mu) \neq \mu$, $\mu^+ \neq \{\mu\}$;
(iv) $\exists \lambda \in \mathbf{B}^n, \Phi^{-1}(\mu) = [\lambda, \mu]$, where $\lambda \neq \mu, \{\mu\} \neq \mu^- = [\lambda, \mu], \Phi(\mu) = \mu$, $\mu^+ = \{\mu\}$.

Proof: (i) We suppose that $(12.3)_{\text{page 131}}$ holds: $\Phi^{-1}(\mu) = \emptyset$. Then $\Phi(\mu) \neq \mu$ and $\mu^+ = [\mu, \Phi(\mu)] \neq [\mu, \mu] = \{\mu\}$. In addition:
(i.1) $\mu^- = \{\mu\}$ is a possibility, otherwise
(i.2) $\mu^- \neq \{\mu\}$. Then $\nu \in \mathbf{B}^n, \nu \neq \mu$ and $\omega \in \mathbf{B}^n$ exist such that $\Phi^\omega(\nu) = \mu$. We denote with $\lambda \in \mathbf{B}^n$ the point that makes true

$$\Phi^{-1}(\Phi(\nu)) = [\lambda, \Phi(\nu)) \text{ or } \Phi^{-1}(\Phi(\nu)) = [\lambda, \Phi(\nu)].$$ [1]

As $\nu, \lambda \in \Phi^{-1}(\Phi(\nu))$ we obviously have $\Phi(\lambda) = \Phi(\nu)$, wherefrom the truth of (12.28). But $\mu \in [\nu, \Phi(\nu)]$; since $\mu \neq \nu$ is a consequence of the initial supposition $\mu^- \neq \{\mu\}$ and because $\mu = \Phi(\nu)$ is impossible, as this would imply $\Phi^{-1}(\mu) \neq \emptyset$, we get $\mu \in (\nu, \Phi(\nu)) \subset (\lambda, \Phi(\lambda))$.

The relations:

$$(\lambda \boxplus \mu) \cap \Phi_\mu = (\lambda \boxplus \mu) \cap (\mu \boxplus \Phi(\mu)) = \emptyset, \tag{12.29}$$

$$\begin{aligned}(\lambda \boxplus \mu) \cup \Phi_\mu &= (\lambda \boxplus \mu) \cup (\mu \boxplus \Phi(\mu)) \\ &= (\lambda \boxplus \mu) \cup (\mu \boxplus \Phi(\lambda)) = \lambda \boxplus \Phi(\lambda) = \Phi_\lambda\end{aligned} \tag{12.30}$$

follow from Theorem 2.3, page 25, thus

$$\begin{aligned}&\forall A \subset \Phi_\mu, (\lambda \boxplus \mu) \cap A = \emptyset, \\ &\forall A \subset \Phi_\mu, (\lambda \boxplus \mu) \Delta A = (\lambda \boxplus \mu) \cup A\end{aligned} \tag{12.31}$$

hold.

---

1 So far $\Phi^{-1}(\Phi(\nu)) = \{\nu\}$ remains a possibility, but this will prove to be false.

At this moment we can compute:

$$[\mu,\Phi(\lambda)]=[\mu,\Phi(\mu)]=\{\mu\oplus\underset{i\in A}{\Xi}\varepsilon^i|A\subset\Phi_\mu\}$$

$$=\{\lambda\oplus\underset{i\in\lambda\boxplus\mu}{\Xi}\varepsilon^i\oplus\underset{i\in A}{\Xi}\varepsilon^i|A\subset\Phi_\mu\}$$

$$=\{\lambda\oplus\underset{i\in(\lambda\boxplus\mu)\Delta A}{\Xi}\varepsilon^i|A\subset\Phi_\mu\}$$

$$\overset{(12.31)}{=}\{\lambda\oplus\underset{i\in(\lambda\boxplus\mu)\cup A}{\Xi}\varepsilon^i|A\subset\Phi_\mu\}$$

$$=\{\lambda\oplus\underset{i\in A}{\Xi}\varepsilon^i|(\lambda\boxplus\mu)\subset A\subset(\lambda\boxplus\mu)\cup\Phi_\mu\}$$

$$\overset{(12.30)}{=}\{\lambda\oplus\underset{i\in A}{\Xi}\varepsilon^i|(\lambda\boxplus\mu)\subset A\subset\Phi_\lambda\}.$$

We prove $[\lambda,\mu]\subset\mu^-$. We take an arbitrary $\nu\in[\lambda,\mu]\subset[\lambda,\Phi(\lambda))$, for which $(\lambda\boxplus\nu)\subset(\lambda\boxplus\mu)$ and $\nu=\lambda\oplus\underset{i\in\lambda\boxplus\nu}{\Xi}\varepsilon^i$. We have $\Phi(\nu)=\Phi(\lambda)$ and let $\omega\in\mathbf{B}^n$. We infer, as

$$(\lambda\boxplus\nu)\cap\Phi_\nu=\emptyset, \tag{12.32}$$

$$(\lambda\boxplus\nu)\cup\Phi_\nu=\Phi_\lambda \tag{12.33}$$

from Theorem 2.3, similarly with (12.29), (12.30) that

$$\forall A\subset\Phi_\nu,(\lambda\boxplus\nu)\cap A=\emptyset,$$

$$\forall A\subset\Phi_\nu,(\lambda\boxplus\nu)\Delta A=(\lambda\boxplus\nu)\cup A. \tag{12.34}$$

In addition:

$$\Phi^\omega(\nu)=\underset{i\in\{1,\ldots,n\}}{\Xi}((1\oplus\omega_i)\nu_i\oplus\omega_i\Phi_i(\nu))\varepsilon^i$$

$$=\underset{i\in\{1,\ldots,n\}}{\Xi}\nu_i\varepsilon^i\oplus\underset{i\in\{1,\ldots,n\}}{\Xi}\omega_i(\nu_i\oplus\Phi_i(\nu))\varepsilon^i$$

$$=\nu\oplus\underset{i\in\{j|j\in\{1,\ldots,n\},\omega_j=1\}\cap\Phi_\nu}{\Xi}\varepsilon^i$$

$$=\lambda\oplus\underset{i\in\lambda\boxplus\nu}{\Xi}\varepsilon^i\oplus\underset{i\in\{j|j\in\{1,\ldots,n\},\omega_j=1\}\cap\Phi_\nu}{\Xi}\varepsilon^i$$

$$=\lambda\oplus\underset{i\in(\lambda\boxplus\nu)\Delta(\{j|j\in\{1,\ldots,n\},\omega_j=1\}\cap\Phi_\nu)}{\Xi}\varepsilon^i$$

$$\overset{(12.34)}{=}\lambda\oplus\underset{i\in(\lambda\boxplus\nu)\cup(\{j|j\in\{1,\ldots,n\},\omega_j=1\}\cap\Phi_\nu)}{\Xi}\varepsilon^i.$$

We define $\omega_j = \begin{cases} 1, & \text{if } j \in \lambda \boxplus \mu, \\ 0, & \text{else} \end{cases}$, $j \in \{1, \dots, n\}$. Then, as far as $(\lambda \boxplus \nu) \subset (\lambda \boxplus \mu) \subset \Phi_\lambda$, we get

$$\begin{aligned}
&(\lambda \boxplus \nu) \cup (\{j | j \in \{1, \dots, n\}, \omega_j = 1\} \cap \Phi_\nu) \\
&= (\lambda \boxplus \nu) \cup ((\lambda \boxplus \mu) \cap \Phi_\nu) \\
&= ((\lambda \boxplus \nu) \cup (\lambda \boxplus \mu)) \cap ((\lambda \boxplus \nu) \cup \Phi_\nu) \overset{(12.33)}{=} (\lambda \boxplus \mu) \cap \Phi_\lambda \\
&= \lambda \boxplus \mu
\end{aligned}$$

therefore

$$\Phi^\omega(\nu) = \lambda \oplus \underset{i \in \lambda \boxplus \mu}{\Xi} \varepsilon^i = \mu.$$

It has resulted that $\nu \in \mu^-$.

We prove $\mu^- \subset [\lambda, \mu]$. We suppose against all reason that $\nu \notin [\lambda, \mu]$ and $\omega \in \mathbf{B}^n$ exist with $\Phi^\omega(\nu) = \mu$. Then $\nu^+ = [\nu, \Phi(\nu)], \mu \neq \nu, \mu \neq \Phi(\nu)$ ($\mu = \Phi(\nu)$ gives the contradiction $\Phi^{-1}(\mu) \neq \emptyset$) imply $\mu \in (\nu, \Phi(\nu))$. The fact that Φ fulfills the generalized tcpo gives $\Phi(\lambda) = \Phi(\mu) = \Phi(\nu)$, hence $\nu \in \Phi^{-1}(\Phi(\lambda))$, therefore $(\nu, \Phi(\nu)) \subset (\lambda, \Phi(\lambda))$. We get in succession $\nu \in [\lambda, \Phi(\lambda)) \setminus [\lambda, \mu]$, $(\lambda \boxplus \nu) \subset \Phi_\lambda$, and $\nu = \lambda \oplus \underset{i \in \lambda \boxplus \nu}{\Xi} \varepsilon^i$. On the other hand, we have not $((\lambda \boxplus \nu) \subset (\lambda \boxplus \mu))$ and we infer like previously:

$$\Phi^\omega(\nu) = \lambda \oplus \underset{i \in (\lambda \boxplus \nu) \cup (\{j | j \in \{1, \dots, n\}, \omega_j = 1\} \cap \Phi_\nu)}{\Xi} \varepsilon^i.$$

The equation $\Phi^\omega(\nu) = \mu$ holds only if the equation

$$(\lambda \boxplus \nu) \cup (\{j | j \in \{1, \dots, n\}, \omega_j = 1\} \cap \Phi_\nu) = \lambda \boxplus \mu$$

holds, i.e. only if $(\lambda \boxplus \nu) \subset (\lambda \boxplus \mu)$. Since we know already that not $((\lambda \boxplus \nu) \subset (\lambda \boxplus \mu))$, we have obtained a contradiction.

(ii) We suppose that $(12.4)_{\text{page } 131}$ holds, i.e. $\Phi^{-1}(\mu) = \{\mu\}$. $\mu \in \mu^-$ and we suppose against all reason the existence of $\nu \in \mathbf{B}^n, \nu \neq \mu$ and $\omega \in \mathbf{B}^n$ such that $\Phi^\omega(\nu) = \mu$. As $\mu \neq \Phi(\nu)$ ($\mu = \Phi(\nu)$ gives the contradiction $\nu \in \Phi^{-1}(\mu)$), we get $\mu \in (\nu, \Phi(\nu))$ hence $\Phi^{-1}(\mu) = \emptyset$, contradiction again. We have proved that $\mu^- = \{\mu\}$. Obviously $\mu^+ = [\mu, \Phi(\mu)] = [\mu, \mu] = \{\mu\}$.

(iii) We suppose now that $(12.5)_{\text{page } 131}$ is true, i.e. $\lambda \in \mathbf{B}^n$ exists, $\lambda \neq \mu$ with $\Phi^{-1}(\mu) = [\lambda, \mu)$, therefore $\Phi(\mu) \neq \mu$ and $\mu^+ = [\mu, \Phi(\mu)] \neq [\mu, \mu] = \{\mu\}$. We must prove that $\mu^- = [\lambda, \mu]$.

$[\lambda, \mu) \subset \mu^-$ is obvious, together with $\mu \in \mu^-$.

In order to prove $\mu^- \subset [\lambda, \mu]$, we suppose against all reason the existence of $\nu \notin [\lambda, \mu]$ and $\omega \in \mathbf{B}^n$ with $\Phi^\omega(\nu) = \mu$, therefore $\mu \in [\nu, \Phi(\nu)]$. Since $\mu \neq \nu$ and $\mu \neq \Phi(\nu)$ ($\mu = \Phi(\nu)$ implies the contradiction $\nu \in \Phi^{-1}(\mu) = [\lambda, \mu)$), we have $\mu \in (\nu, \Phi(\nu))$, i.e. $\Phi^{-1}(\mu) = \emptyset$, contradiction.

(iv) We have the truth of $(12.6)_{\text{page } 131}$, i.e. $\lambda \in \mathbf{B}^n$ exists, $\lambda \neq \mu$ such that $\Phi^{-1}(\mu) = [\lambda, \mu]$. We infer $\Phi(\mu) = \mu$ and $\mu^+ = [\mu, \Phi(\mu)] = [\mu, \mu] = \{\mu\}$. The fact that $\mu^- = [\lambda, \mu]$ is proved like at (iii) □

12.5 Asynchronous and Synchronous Transient Points

Definition 12.2 A transient point $\mu \in \mathbf{B}^n$ with the property that $\exists \lambda \in \mathbf{B}^n, \mu \in (\lambda, \Phi(\lambda))$ is called **asynchronous,** and a transient point μ with the property that $\exists \lambda \in \mathbf{B}^n, \Phi^{-1}(\mu) = [\lambda, \mu)$ is called **synchronous.**

Remark 12.5 The previous definition is a consequence of Theorem 12.6, where the asynchronous transient points have occurred at (i.2) and the synchronous transient points have occurred at (iii). Since (i)–(iv) cover all the possibilities, we conclude that in the case of the strong generalized tcpo, all the transient points are either synchronous or asynchronous. If Φ fulfills the strong tcpo, it fulfills the strong generalized tcpo also and its transient points are synchronous.

If Φ fulfills the generalized tcpo, but not the strong generalized tcpo, see Figure 11.3, page 120, then Definition 12.2 has no sense, as far as $(0, 0, 1)$ is synchronous if it ends a computation started in $(1, 0, 1)$ and asynchronous if it is an intermediate value in a computation started in $(0, 1, 1)$.

12.6 The Sets of Predecessors and Successors

Remark 12.6 For any $\mu \in \mathbf{B}^n$, the properties

$$\mu^- \supset \{\mu\} \cup \Phi^{-1}(\mu),$$

$$\mu^+ \supset \{\mu\} \cup \{\Phi(\mu)\}$$

are still true, see Remark 11.6, page 121. In fact, even the example that was given there refers to Figure 11.1, page 119 where the strong generalized tcpo is fulfilled.

Theorem 12.7 If Φ fulfills the strong generalized tcpo then for any $\mu \in \mathbf{B}^n$:

(a) a unique $\lambda \in \mathbf{B}^n$ exists such that

$$\mu^- = [\lambda, \mu]$$

and in addition if $(\lambda, \mu) \neq \emptyset$ then for any $\nu \in (\lambda, \mu)$ we get $\nu^- = [\lambda, \nu]$;

(b) we have

$$O^-(\mu) = [\mu, \lambda] \cup \Phi^{-1}(\lambda) \cup \Phi^{-1}(\Phi^{-1}(\lambda)) \cup \ldots \tag{12.35}$$

$$O^+(\mu) = [\mu, \Phi(\mu)] \cup [\Phi(\mu), \Phi^{(2)}(\mu)] \cup \ldots \tag{12.36}$$

Proof: (a) This is a consequence of Theorem 12.6, page 137.

(b) We prove equation (12.35) by referring to the possibilities (i)–(iv) from Theorem 12.6.

Case (i.1) μ is a source, when $\Phi^{-1}(\mu) = \emptyset, \mu^- = \{\mu\}$

$$O^-(\mu) \overset{(1.29)_{\text{page 16}}}{=} \{\mu\} \cup \{\mu\} \cup \{\mu\} \cup \ldots = \{\mu\}$$

and on the other hand (12.35) written for $\lambda = \mu$ gives

$$O^-(\mu) = \{\mu\} \cup \Phi^{-1}(\mu) \cup \Phi^{-1}(\Phi^{-1}(\mu)) \cup \ldots = \{\mu\}.$$

Statement (12.35) is proved.

Case (ii) μ is an isolated fixed point, $\Phi^{-1}(\mu) = \{\mu\}, \mu^- = \{\mu\}$

$$O^-(\mu) \overset{(1.29)_{\text{page 16}}}{=} \{\mu\} \cup \{\mu\} \cup \{\mu\} \cup \ldots = \{\mu\},$$

$$O^-(\mu) \overset{(12.35)}{=} \{\mu\} \cup \{\mu\} \cup \{\mu\} \cup \ldots = \{\mu\}$$

and statement (12.35) is proved again.

The following three cases are treated together:

Case (i.2) μ is an asynchronous transient point, $\Phi^{-1}(\mu) = \emptyset$, $\exists \lambda \in \mathbf{B}^n$, $(\Phi^{-1}(\Phi(\lambda)) = [\lambda, \Phi(\lambda))$ or $\Phi^{-1}(\Phi(\lambda)) = [\lambda, \Phi(\lambda)])$, $\mu \in (\lambda, \Phi(\lambda))$, $\mu^- = [\lambda, \mu]$,

Case (iii) μ is a synchronous transient point, $\exists \lambda \in \mathbf{B}^n, \Phi^{-1}(\mu) = [\lambda, \mu)$, $\mu^- = [\lambda, \mu]$,

Case (iv) μ is a sink, $\exists \lambda \in \mathbf{B}^n, \Phi^{-1}(\mu) = [\lambda, \mu], \mu^- = [\lambda, \mu]$.

In (i.2) we have

$$\forall \nu \in [\lambda, \mu], \Phi(\lambda) = \Phi(\mu) = \Phi(\nu), \text{ thus } \nu \in \lambda^+ \supsetneq \{\lambda\},$$

in (iii) we have

$$\forall \nu \in [\lambda, \mu), \Phi(\lambda) = \Phi(\nu), \text{ thus } \nu \in \lambda^+ \supsetneq \{\lambda\},$$

and (iv) implies

$$\forall \nu \in [\lambda, \mu], \Phi(\lambda) = \Phi(\mu) = \Phi(\nu) = \mu, \text{ thus } \nu \in \lambda^+ \supsetneq \{\lambda\}.$$

We have obtained that λ is not a sink and it is not an isolated fixed point either, since these would mean that $\lambda^+ = \{\lambda\}$. We prove now that it is not an asynchronous transient point, by supposing against all reason that $\Phi^{-1}(\lambda) = \emptyset$ and $\lambda' \in \mathbf{B}^n$ exists such that $(\Phi^{-1}(\Phi(\lambda')) = [\lambda', \Phi(\lambda'))$ or $\Phi^{-1}(\Phi(\lambda')) = [\lambda', \Phi(\lambda')])$ and $\lambda \in (\lambda', \Phi(\lambda'))$. We infer $\Phi(\lambda) = \Phi(\lambda')$, therefore

$$[\lambda, \Phi(\lambda)) = \Phi^{-1}(\Phi(\lambda)) = \Phi^{-1}(\Phi(\lambda')) = [\lambda', \Phi(\lambda')) = [\lambda', \Phi(\lambda))$$

if $\Phi(\Phi(\lambda)) \neq \Phi(\lambda)$, and

$$[\lambda, \Phi(\lambda)] = \Phi^{-1}(\Phi(\lambda)) = \Phi^{-1}(\Phi(\lambda')) = [\lambda', \Phi(\lambda')] = [\lambda', \Phi(\lambda)]$$

if $\Phi(\Phi(\lambda)) = \Phi(\lambda)$. In both situations, we get $\lambda = \lambda'$, contradiction showing that λ is not an asynchronous transient point.

The conclusion is that the following possibilities occur.

Case (j) A sequence $\lambda^1, \lambda^2, \lambda^3, \ldots \in \mathbf{B}^n$ of synchronous transient points exists, with $\Phi^{-1}(\lambda) = [\lambda^1, \lambda), \Phi^{-1}(\lambda^1) = [\lambda^2, \lambda^1), \Phi^{-1}(\lambda^2) = [\lambda^3, \lambda^2), \ldots$

Case (jj) Either λ is a source, $\Phi^{-1}(\lambda) = \emptyset, \lambda^- = \{\lambda\}$, or a finite sequence $\lambda^1, \lambda^2, \ldots, \lambda^p \in \mathbf{B}^n$ of synchronous transient points exists, with $\Phi^{-1}(\lambda) = [\lambda^1, \lambda), \ldots, \Phi^{-1}(\lambda^p) = [\lambda^{p+1}, \lambda^p)$ and $\lambda^{p+1} \in \mathbf{B}^n$ is a source, $\Phi^{-1}(\lambda^{p+1}) = \emptyset, \lambda^{p+1-} = \{\lambda^{p+1}\}$.

In Case (j):

$$O^-(\mu) \overset{(1.29)_{\text{page 16}}}{=} [\mu, \lambda] \cup [\lambda, \lambda^1] \cup [\lambda^1, \lambda^2] \cup \ldots,$$

$$O^-(\mu) \overset{(12.35)}{=} [\mu, \lambda] \cup [\lambda^1, \lambda) \cup [\lambda^2, \lambda^1) \cup \ldots$$

and statement (12.35) is proved. In Case (jj) we infer if λ is a source that

$$O^-(\mu) \overset{(1.29)_{\text{page 16}}}{=} [\mu, \lambda] \cup \{\lambda\} = [\mu, \lambda],$$

$$O^-(\mu) \overset{(12.35)}{=} [\mu, \lambda] \cup \Phi^{-1}(\lambda) = [\mu, \lambda]$$

proving statement (12.35), and if $\lambda^1, \lambda^2, \ldots, \lambda^p \in \mathbf{B}^n$ are synchronous transient points with $\lambda^{p+1} \in \mathbf{B}^n$ source, we conclude that

$$O^-(\mu) \overset{(1.29)_{\text{page 16}}}{=} [\mu, \lambda] \cup [\lambda, \lambda^1] \cup \ldots \cup [\lambda^p, \lambda^{p+1}] \cup \{\lambda^{p+1}\},$$

$$O^-(\mu) \overset{(12.35)}{=} [\mu, \lambda] \cup [\lambda^1, \lambda) \cup \ldots \cup [\lambda^{p+1}, \lambda^p)$$

proving statement (12.35) again. We have used in the last equations the fact that

$$\begin{aligned}
\bigcup_{\omega \in \mu^-} \omega^- &= \bigcup_{\omega \in [\lambda, \mu]} \omega^- = \bigcup_{\omega \in \{\lambda\} \cup (\lambda, \mu) \cup \{\mu\}} \omega^- \\
&= \bigcup_{\omega \in \{\lambda\}} \omega^- \cup \bigcup_{\omega \in (\lambda, \mu)} \omega^- \cup \bigcup_{\omega \in \{\mu\}} \omega^- \\
&= \lambda^- \overset{(a)}{\cup} \bigcup_{\omega \in (\lambda, \mu)} [\lambda, \omega] \cup [\lambda, \mu] = \lambda^- \cup [\lambda, \mu] = \lambda^- \cup \mu^- \\
&= [\lambda, \lambda^1] \cup [\mu, \lambda],
\end{aligned}$$

$$\begin{aligned}
\bigcup_{\omega \in \bigcup_{\delta \in \mu^-} \delta^-} \omega^- &= \bigcup_{\omega \in [\lambda, \lambda^1] \cup [\mu, \lambda]} \omega^- = \bigcup_{\omega \in [\lambda, \lambda^1]} \omega^- \cup \bigcup_{\omega \in [\mu, \lambda]} \omega^- \\
&= (\lambda^- \cup \lambda^{1-}) \cup (\mu^- \cup \lambda^-)
\end{aligned}$$

$$= \lambda^{1-} \cup \lambda^{-} \cup \mu^{-} = [\lambda^1, \lambda^2] \cup [\lambda, \lambda^1] \cup [\mu, \lambda],$$

$$\Phi^{-1}(\Phi^{-1}(\lambda)) = \Phi^{-1}([\lambda^1, \lambda)) = \bigcup_{\omega \in [\lambda^1, \lambda)} \Phi^{-1}(\omega) = \bigcup_{\omega \in \{\lambda^1\} \cup (\lambda^1, \lambda)} \Phi^{-1}(\omega)$$

$$= \Phi^{-1}(\lambda^1) \cup \bigcup_{\omega \in (\lambda^1, \lambda)} \Phi^{-1}(\omega) = [\lambda^2, \lambda^1) \cup \emptyset = [\lambda^2, \lambda^1),$$

etc.

Equation (12.36) coincides with (11.29) $_{\text{page 124}}$ □

12.7 Isomorphisms vs the Strong Generalized tcpo

Theorem 12.8 Let the functions $\Phi, \Psi : \mathbf{B}^n \longrightarrow \mathbf{B}^n$ and the isomorphism $(h, h') \in Iso(\Phi, \Psi)$. We suppose that $h \in Af(\mathbf{B}^n)$ and that Φ fulfills the strong generalized tcpo. In these circumstances Ψ fulfills the strong generalized tcpo.

Proof: We refer to the version (12.15) and ((12.3 or …or (12.6)) of Definition 12.1, page 133 of the strong generalized tcpo.

We take an arbitrary $\nu \in \mathbf{B}^n$ and the hypothesis states the commutativity of the diagram

$$\begin{array}{ccc} \mathbf{B}^n & \xrightarrow{\Phi^\nu} & \mathbf{B}^n \\ h \downarrow & & \downarrow h \\ \mathbf{B}^n & \xrightarrow{\Psi^{h'(\nu)}} & \mathbf{B}^n \end{array}$$

As Φ fulfills the strong generalized tcpo, we get from Theorem 12.5, page 136 that Φ^ν fulfills the strong generalized tcpo too. We show for the beginning that

$$\forall \mu' \in \mathbf{B}^n, \forall \lambda' \in (\mu', \Psi^{h'(\nu)}(\mu')), (\Psi^{h'(\nu)})^{-1}(\lambda') = \emptyset \tag{12.37}$$

and we suppose against all reason that this is not true. We infer the existence of $\mu' \in \mathbf{B}^n$, $\lambda' \in (\mu', \Psi^{h'(\nu)}(\mu'))$ and $\omega' \in \mathbf{B}^n$ such that $\Psi^{h'(\nu)}(\omega') = \lambda'$. We denote $\mu = h^{-1}(\mu')$, $\lambda = h^{-1}(\lambda')$ and $\omega = h^{-1}(\omega')$. We have:

$$\lambda' = \Psi^{h'(\nu)}(\omega') = \Psi^{h'(\nu)}(h(\omega)) = h(\Phi^\nu(\omega)),$$

thus $\lambda = h^{-1}(\lambda') = \Phi^\nu(\omega)$. On the other hand, the fact that $h \in Af(\mathbf{B}^n)$ implies $h^{-1} \in Af(\mathbf{B}^n)$, therefore (see Theorem 2.6, page 28)

$$\lambda \in h^{-1}(\mu', \Psi^{h'(\nu)}(\mu')) = (h^{-1}(\mu'), h^{-1}(\Psi^{h'(\nu)}(\mu')))$$
$$= (\mu, h^{-1}(\Psi^{h'(\nu)}(h(\mu)))) = (\mu, h^{-1}(h(\Phi^\nu(\mu))))$$
$$= (\mu, \Phi^\nu(\mu)).$$

A contradiction has resulted with the request $(\Phi^\nu)^{-1}(\lambda) = \emptyset$, hence (12.37) takes place.

We take now $v \in \mathbf{B}^n$, $\mu' \in \mathbf{B}^n$ arbitrary, fixed. We denote $\mu = h^{-1}(\mu')$ and we obtain the next possibilities, which are inferred by rewriting (12.3)–(12.6) from page 131 for Φ^v.

Case (1)

$$(\Phi^v)^{-1}(\mu) = \emptyset \tag{12.38}$$

and we prove the fulfillment of

$$(\Psi^{h'(v)})^{-1}(\mu') = \emptyset. \tag{12.39}$$

We suppose against all reason that (12.39) is false. This gives the existence of $\mu'' \in \mathbf{B}^n$ with $\Psi^{h'(v)}(\mu'') = \mu'$. With the notation $\tilde{\mu} = h^{-1}(\mu'')$, we can write:

$$\mu' = \Psi^{h'(v)}(\mu'') = \Psi^{h'(v)}(h(\tilde{\mu})) = h(\Phi^v(\tilde{\mu})) = h(\mu).$$

As h is bijection, we get $\Phi^v(\tilde{\mu}) = \mu$, contradiction with (12.38). Statement (12.39) holds.

Case (2)

$$(\Phi^v)^{-1}(\mu) = \{\mu\} \tag{12.40}$$

and, for $\mu' = h(\mu)$, we prove the truth of

$$(\Psi^{h'(v)})^{-1}(\mu') = \{\mu'\}. \tag{12.41}$$

$(\Psi^{h'(v)})^{-1}(\mu') \subset \{\mu'\}$. We suppose against all reason that the inclusion does not hold, thus $\mu'' \neq \mu'$ exists with the property $\Psi^{h'(v)}(\mu'') = \mu'$. We put $\tilde{\mu} = h^{-1}(\mu'')$ and we have:

$$\mu' = \Psi^{h'(v)}(\mu'') = \Psi^{h'(v)}(h(\tilde{\mu})) = h(\Phi^v(\tilde{\mu})) = h(\mu),$$

therefore

$$\Phi^v(\tilde{\mu}) = \mu. \tag{12.42}$$

But $\mu'' \neq \mu'$ implies $\tilde{\mu} = h^{-1}(\mu'') \neq h^{-1}(\mu') = \mu$ and (12.42) is in contradiction with (12.40).

$\{\mu'\} \subset (\Psi^{h'(v)})^{-1}(\mu')$. We have

$$\Psi^{h'(v)}(\mu') = \Psi^{h'(v)}(h(\mu)) = h(\Phi^v(\mu)) = h(\mu) = \mu'.$$

Statement (12.41) holds.

Case (3) $\exists \lambda \in \mathbf{B}^n$ such that

$$(\Phi^v)^{-1}(\mu) = [\lambda, \mu) \tag{12.43}$$

and we prove that

$$(\Psi^{h'(v)})^{-1}(\mu') = [\lambda', \mu') \tag{12.44}$$

with the notations $\lambda' = h(\lambda)$, $\mu' = h(\mu)$.

$(\Psi^{h'(v)})^{-1}(\mu') \subset [\lambda', \mu')$. Let us take $\omega' \in (\Psi^{h'(v)})^{-1}(\mu')$ arbitrary, therefore we get $\Psi^{h'(v)}(\omega') = \mu'$. We use the notation $\omega = h^{-1}(\omega')$. We obtain:

$$\mu' = \Psi^{h'(v)}(\omega') = \Psi^{h'(v)}(h(\omega)) = h(\Phi^v(\omega)),$$

wherefrom $\Phi^v(\omega) = h^{-1}(\mu') = \mu$ and $\omega \overset{(12.43)}{\in} [\lambda, \mu)$. We infer:

$$\omega' = h(\omega) \in h([\lambda, \mu)) = [h(\lambda), h(\mu)) = [\lambda', \mu').$$

$[\lambda', \mu') \subset (\Psi^{h'(v)})^{-1}(\mu')$. We take $\omega' \in [\lambda', \mu')$ arbitrary and we denote $\omega = h^{-1}(\omega')$ like before. We have:

$$\omega \in h^{-1}([\lambda', \mu')) = [h^{-1}(\lambda'), h^{-1}(\mu')) = [\lambda, \mu),$$

$$\Psi^{h'(v)}(\omega') = \Psi^{h'(v)}(h(\omega)) = h(\Phi^v(\omega)) = h(\mu) = \mu',$$

in other words $\omega' \in (\Psi^{h'(v)})^{-1}(\mu')$. Statement (12.44) is proved.

Case (4) $\exists \lambda \in \mathbf{B}^n$ such that

$$(\Phi^v)^{-1}(\mu) = [\lambda, \mu]. \tag{12.45}$$

Proving that

$$(\Psi^{h'(v)})^{-1}(\mu') = [\lambda', \mu'], \tag{12.46}$$

where $\lambda' = h(\lambda)$, is similar with the proof of Case (3).

The conclusion is that $\Psi^{h'(v)}$ satisfies the strong generalized tcpo. As v was arbitrary and h' is bijection, we obtain from Theorem 12.5 that Ψ fulfills the strong generalized tcpo □

12.8 Antiisomorphisms vs the Strong Generalized tcpo

Theorem 12.9 The functions $\Phi, \Psi : \mathbf{B}^n \longrightarrow \mathbf{B}^n$ and the antiisomorphism $(h, h')^{\smallfrown} \in Iso^{\smallfrown}(\Phi, \Psi)$ are given. If Φ fulfills the strong generalized tcpo, $h \in Af(\mathbf{B}^n)$ and

$$\forall v \in \mathbf{B}^n, \Phi^v(\mathbf{B}^n) = \mathbf{B}^n \tag{12.47}$$

is true, then Ψ satisfies the strong generalized tcpo and also the invariance

$$\forall v \in \mathbf{B}^n, \Psi^v(\mathbf{B}^n) = \mathbf{B}^n. \tag{12.48}$$

Proof: The hypothesis of Theorem 11.10, page 128 is fulfilled, wherefrom we get that (12.48) is true and Ψ fulfills the generalized tcpo too. But Ψ is bijective, and Theorem 12.4, page 135 shows that it satisfies also the strong generalized tcpo □

13

Time-Reversal Symmetry

If we make a film and the film is watched backward, from the end to the beginning, we have an example of time-reversal.

In music, the reverse tape effects are special effects created by recording sound onto magnetic tape and then physically reversing the tape so that when the tape is played back, the sounds recorded on it are heard in reverse.

The survey [20] gives connections of time-reversal symmetry with physics. In classical mechanics, the classical ideal pendulum without loss of energy due to friction accepts time-reversal, but in the presence of friction this possibility disappears. In thermodynamics, Loschmidt refers to the time-reversal symmetry of the microscopic equations of motion of a macroscopic number of gas molecules, representing a contradiction with the Boltzmann's second law of thermodynamics (the Loschmidt's paradox). And in quantum mechanics, Wiegner introduced in 1930 a quantum mechanical version of the classical conventional time-reversal operator. He explains this way the twofold degeneracy of energy levels in systems with an odd number of electrons in the absence of an electric field.

In dynamical systems [20], time-reversal symmetry was used for the first time in 1915 by Birkhoff, in the 1960s it was studied by mathematicians such as DeVogelaere, Heinbockel, Struble, Moser, Bibikov, Pliss, Hale, and later in the 1970s–1980s Devaney, Arnol'd, Sevryuk had their own contributions.

We define the time-reversal symmetry of Φ and Ψ by the fact that for each $\mu \in \mathbf{B}^n$, the immediate predecessors of one function coincide with the immediate successors of the other one and vice versa.

We show that the symmetrical function of a function Φ is unique.

The action of the isomorphisms/antiisomorphisms of the time-reversed symmetrical functions on predecessors and successors is studied/suggested.

Boolean Functions: Topics in Asynchronicity, First Edition. Serban E. Vlad.

13.1 Definition

Theorem 13.1 Let $\Phi, \Psi : \mathbf{B}^n \to \mathbf{B}^n$. The conjunction of the statements

$$\mu_{\Phi}^{-} = \mu_{\Psi}^{+}, \tag{13.1}$$

$$\mu_{\Psi}^{-} = \mu_{\Phi}^{+}, \tag{13.2}$$

where $\mu \in \mathbf{B}^n$ is equivalent with the conjunction of the statements

$$\forall \nu \in \mathbf{B}^n, \exists \lambda \in \mathbf{B}^n, (\Phi^{\lambda} \circ \Psi^{\nu})(\mu) = \mu, \tag{13.3}$$

$$\forall \lambda \in \mathbf{B}^n, \exists \nu \in \mathbf{B}^n, (\Psi^{\nu} \circ \Phi^{\lambda})(\mu) = \mu, \tag{13.4}$$

where $\mu \in \mathbf{B}^n$.

Proof: In order to prove that $\forall \mu \in \mathbf{B}^n$, ((13.1) and (13.2)) is equivalent with $\forall \mu \in \mathbf{B}^n$, ((13.3) and (13.4)), we fix an arbitrary μ. Here is the proof of this equivalence.

(13.1)$\Longrightarrow$(13.3). Let $\nu \in \mathbf{B}^n$ and $\Psi^{\nu}(\mu) = \mu' \in \mu_{\Psi}^{+}$. As $\mu' \in \mu_{\Phi}^{-}$, $\lambda \in \mathbf{B}^n$ exists with $\Phi^{\lambda}(\mu') = \mu$, thus (13.3) holds.

(13.2)$\Longrightarrow$(13.4). For $\lambda \in \mathbf{B}^n$, we have $\Phi^{\lambda}(\mu) = \mu' \in \mu_{\Phi}^{+}$. As $\mu' \in \mu_{\Psi}^{-}$, some $\nu \in \mathbf{B}^n$ exists with $\Psi^{\nu}(\mu') = \mu$, showing the truth of (13.4).

(13.4)$\Longrightarrow$ $\mu_{\Phi}^{-} \subset \mu_{\Psi}^{+}$. Let an arbitrary $\mu' \in \mu_{\Phi}^{-}$, then $\lambda \in \mathbf{B}^n$ exists such that $\Phi^{\lambda}(\mu') = \mu$. From (13.4), we have the existence of $\nu \in \mathbf{B}^n$ with $(\Psi^{\nu} \circ \Phi^{\lambda})(\mu') = \mu' = \Psi^{\nu}(\mu)$, thus $\mu' \in \mu_{\Psi}^{+}$.

(13.3)$\Longrightarrow$ $\mu_{\Psi}^{+} \subset \mu_{\Phi}^{-}$. We take an arbitrary $\mu' \in \mu_{\Psi}^{+}$, for which $\nu \in \mathbf{B}^n$ exists with $\mu' = \Psi^{\nu}(\mu)$. Relation (13.3) shows the existence of $\lambda \in \mathbf{B}^n$ for which $(\Phi^{\lambda} \circ \Psi^{\nu})(\mu) = \mu = \Phi^{\lambda}(\mu')$, meaning that $\mu' \in \mu_{\Phi}^{-}$.

(13.3)$\Longrightarrow$ $\mu_{\Psi}^{-} \subset \mu_{\Phi}^{+}$. For an arbitrary $\mu' \in \mu_{\Psi}^{-}$, some $\nu \in \mathbf{B}^n$ exists with $\Psi^{\nu}(\mu') = \mu$. Relation (13.3) shows the existence of $\lambda \in \mathbf{B}^n$ with $(\Phi^{\lambda} \circ \Psi^{\nu})(\mu') = \mu' = \Phi^{\lambda}(\mu)$, thus $\mu' \in \mu_{\Phi}^{+}$.

(13.4)$\Longrightarrow$ $\mu_{\Phi}^{+} \subset \mu_{\Psi}^{-}$. Let $\mu' \in \mu_{\Phi}^{+}$ arbitrary, thus $\lambda \in \mathbf{B}^n$ exists with $\mu' = \Phi^{\lambda}(\mu)$. From (13.4) we have the existence of $\nu \in \mathbf{B}^n$ such that $(\Psi^{\nu} \circ \Phi^{\lambda})(\mu) = \mu = \Psi^{\nu}(\mu')$, giving $\mu' \in \mu_{\Psi}^{-}$ □

Definition 13.1 If one of the properties

$\forall \mu \in \mathbf{B}^n$, ((13.1) and (13.2)),

$\forall \mu \in \mathbf{B}^n$, ((13.3) and (13.4)),

is fulfilled, we say that the **time-reversal symmetry** of the functions Φ and Ψ holds.

Remark 13.1 Several types of symmetry and antisymmetry of the Boolean functions exist, for example, we have defined at page 39, Definitions 3.3 and 3.4 the symmetry relative to translations and at page 51, Definitions 4.4 and 4.5

the antisymmetry relative to translations. The reasons of calling the previous symmetry "time-reversal" become clear after introducing the flows and this is beyond the purpose of the present work.

Remark 13.2 We notice the suggestion given by (13.1) and (13.2) that Φ, Ψ inverse causality while (13.3) and (13.4) seem to present Φ, Ψ as "inverse" to each other.

Remark 13.3 The condition of time-reversal symmetry of Φ, Ψ does not mean the existence of two antimorphisms $(1_{\mathbf{B}^n}, h') : \Psi \to \Phi, (1_{\mathbf{B}^n}, g') : \Phi \to \Psi$ with $\lambda = h'(\nu), \nu = g'(\lambda)$ because the values of λ in (13.3) and ν in (13.4) depend on μ.

Theorem 13.2 The time-reversal symmetry of Φ and Ψ holds if and only if the time-reversal symmetry of Φ^* and Ψ^* holds.

Proof: We use Theorem 1.11, page 17.

Only if. We take some arbitrary $\mu \in \mathbf{B}^n$ and we have:

$$\overline{\mu_{\Phi^*}^-} = \overline{\mu_{\Phi}^-} = \overline{\mu_{\Psi}^+} = \overline{\mu_{\Psi^*}^+},$$

$$\overline{\mu_{\Phi^*}^+} = \overline{\mu_{\Phi}^+} = \overline{\mu_{\Psi}^-} = \overline{\mu_{\Psi^*}^-},$$

etc □

Theorem 13.3 If Φ, Ψ are time-reverse symmetrical, then $\forall \mu \in \mathbf{B}^n$,

$$O_{\Phi}^-(\mu) = O_{\Psi}^+(\mu), \tag{13.5}$$

$$O_{\Phi}^+(\mu) = O_{\Psi}^-(\mu). \tag{13.6}$$

Proof: (13.5) results from $(1.29)_{page\ 16}$ and $(1.30\ )_{page\ 16}$:

$$O_{\Phi}^-(\mu) = \mu_{\Phi}^- \cup \bigcup_{\lambda \in \mu_{\Phi}^-} \lambda_{\Phi}^- \cup \bigcup_{\delta \in \bigcup\limits_{\lambda \in \mu_{\Phi}^-} \lambda_{\Phi}^-} \delta_{\Phi}^- \cup \ldots$$

$$= \mu_{\Psi}^+ \cup \bigcup_{\lambda \in \mu_{\Psi}^+} \lambda_{\Psi}^+ \cup \bigcup_{\delta \in \bigcup\limits_{\lambda \in \mu_{\Psi}^+} \lambda_{\Psi}^+} \delta_{\Psi}^+ \cup \ldots = O_{\Psi}^+(\mu)$$

□

Remark 13.4 The condition $\forall \mu \in \mathbf{B}^n$, ((13.5) and (13.6)) is necessary for the time-reversal symmetry of Φ, Ψ. Is it also sufficient?

13.2 Examples

Example 13.1 The identity $1_{\mathbf{B}^n} : \mathbf{B}^n \to \mathbf{B}^n$ is the time-reversed symmetrical function of itself; we get $\forall \mu \in \mathbf{B}^n, \mu^+_{1_{\mathbf{B}^n}} = \mu^-_{1_{\mathbf{B}^n}} = \{\mu\}$.

Example 13.2 Let $\omega \in \mathbf{B}^n$. We define the constant functions $\Phi, \Psi : \mathbf{B}^n \to \mathbf{B}^n$ in the following way: $\forall \mu \in \mathbf{B}^n, \Phi(\mu) = \omega$ and $\forall \mu \in \mathbf{B}^n, \Psi(\mu) = \overline{\omega} = (\overline{\omega}_1, \ldots, \overline{\omega}_n)$. We have:

$$\overline{\omega}^-_\Phi = \{\overline{\omega}\} = \overline{\omega}^+_\Psi, \quad \overline{\omega}^+_\Phi = \mathbf{B}^n = \overline{\omega}^-_\Psi,$$
$$\omega^-_\Phi = \mathbf{B}^n = \omega^+_\Psi, \quad \omega^+_\Phi = \{\omega\} = \omega^-_\Psi,$$
$$\forall \mu \in (\overline{\omega}, \omega), \quad \mu^-_\Phi = [\overline{\omega}, \mu] = \mu^+_\Psi, \quad \mu^+_\Phi = [\mu, \omega] = \mu^-_\Psi.$$

We give the example of the functions $\Phi, \Psi : \mathbf{B}^2 \to \mathbf{B}^2$ from Figure 13.1 (a), (b) which are constant: $\forall \mu \in \mathbf{B}^2$,

$$\Phi(\mu) = (1, 1),$$
$$\Psi(\mu) = (0, 0).$$

Notice the existence of the same arrows at (a), (b), with different senses however, that show the meaning of time-reversal symmetry.

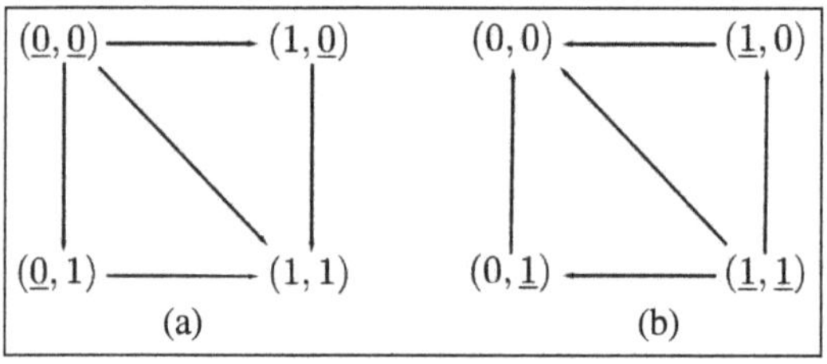

Figure 13.1 The time-reversal symmetry of two constant functions.

Example 13.3 The functions $\Phi, \Psi : \mathbf{B}^3 \to \mathbf{B}^3$, $\forall \mu \in \mathbf{B}^3, \Phi(\mu) = (\mu_1, \overline{\mu_1} \cup \mu_1\mu_2, \overline{\mu_1} \cup \mu_1\mu_3)$ and $\forall \mu \in \mathbf{B}^3, \Psi(\mu) = (\mu_1, \mu_1\mu_2, \mu_1\mu_3)$ are time-reversed symmetrical. We have:

$$\forall \mu \in \{(1,0,0), (1,0,1), (1,1,0), (1,1,1)\}, \mu^-_\Phi = \mu^+_\Psi = \{\mu\},$$
$$\mu^+_\Phi = \mu^-_\Psi = \{\mu\},$$
$$(0,0,0)^+_\Phi = (0,0,0)^-_\Psi = \{(0,0,0), (0,1,0), (0,0,1), (0,1,1)\} \text{ etc.}$$

The state portraits of Φ and Ψ have been drawn in Figures 13.2 and 13.3.

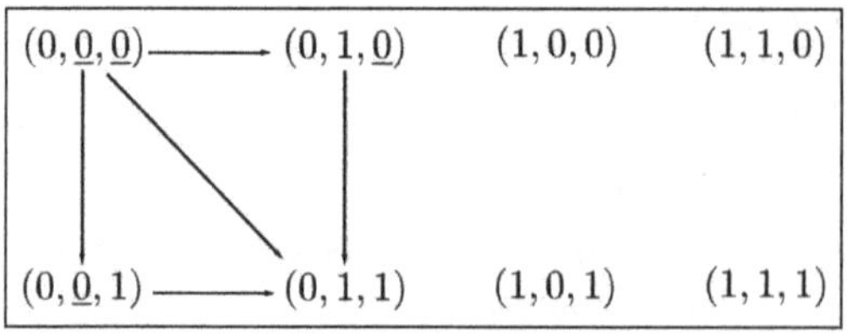

Figure 13.2 The function $\Phi(\mu_1, \mu_2, \mu_3) = (\mu_1, \overline{\mu_1} \cup \mu_1\mu_2, \overline{\mu_1} \cup \mu_1\mu_3)$.

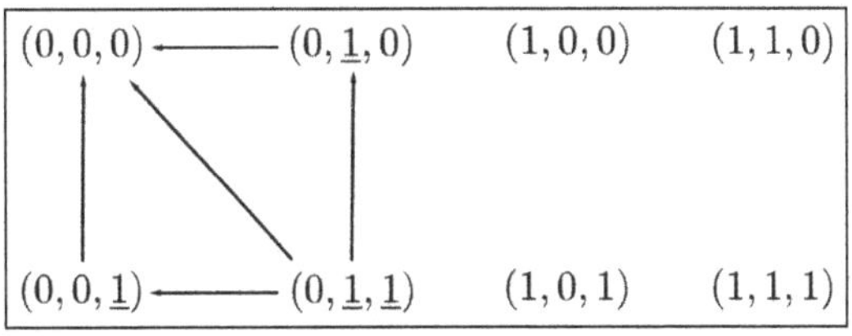

Figure 13.3 The function $\Psi(\mu_1, \mu_2, \mu_3) = (\mu_1, \mu_1\mu_2, \mu_1\mu_3)$.

13.3 The Uniqueness of the Symmetrical Function

Theorem 13.4 Let the functions $\Phi, \Psi, \Gamma : \mathbf{B}^n \to \mathbf{B}^n$. The time-reversal symmetry of Φ and Ψ, together with the time-reversal symmetry of Φ and Γ imply $\Psi = \Gamma$.

Proof: We suppose against all reason the contrary, that $\Psi \neq \Gamma$, meaning the existence of $\mu \in \mathbf{B}^n$ with $\Psi(\mu) \neq \Gamma(\mu)$. We infer the existence of the sets $I, J \subset \{1, \dots, n\}$ with the property that $\Psi(\mu) = \mu \oplus \underset{i\in I}{\Xi} \varepsilon^i$, $\Gamma(\mu) = \mu \oplus \underset{i\in J}{\Xi} \varepsilon^i$ and $I \neq J$. Without loss of generality, we can suppose the existence of some $i \in I \setminus J$. We infer that $\mu \oplus \varepsilon^i \in [\mu, \mu \oplus \underset{j\in I}{\Xi} \varepsilon^j] = \mu_\Psi^+$, $\mu \oplus \varepsilon^i \notin [\mu, \mu \oplus \underset{j\in J}{\Xi} \varepsilon^j] = \mu_\Gamma^+$, and this represents a contradiction with the hypothesis stating that $\mu_\Psi^+ = \mu_\Phi^- = \mu_\Gamma^+$ □

13.4 Isomorphisms and Antiisomorphisms vs Time-Reversal Symmetry

Theorem 13.5 We consider the functions $\Gamma, \Phi, \Psi, \Upsilon : \mathbf{B}^n \to \mathbf{B}^n$.

(a) If $(h, h') \in Iso(\Gamma, \Phi)$, $(g, g') \in Iso(\Psi, \Upsilon)$ and Φ and Ψ are time-reversed symmetrical then $\forall \mu \in \mathbf{B}^n$,

$$g(h(\mu_\Gamma^-)) = g(h(\mu))_\Upsilon^+, \tag{13.7}$$

$$g(h(\mu_\Gamma^+)) = g(h(\mu))_\Upsilon^-, \tag{13.8}$$

$$g(h(O_\Gamma^-(\mu))) = O_\Upsilon^+(g(h(\mu))), \tag{13.9}$$

$$g(h(O_\Gamma^+(\mu))) = O_\Upsilon^-(g(h(\mu))). \tag{13.10}$$

(b) We suppose that Γ and Φ are time-reversed symmetrical, $(h, h') \in Iso(\Phi, \Psi)$, while Ψ and Υ are time-reversed symmetrical too. Then $\forall \mu \in \mathbf{B}^n$,

$$h(\mu_\Gamma^-) = h(\mu)_\Upsilon^-, \tag{13.11}$$

$$h(\mu_\Gamma^+) = h(\mu)_\Upsilon^+, \tag{13.12}$$

$$h(O_\Gamma^-(\mu)) = O_\Upsilon^-(h(\mu)), \tag{13.13}$$

$$h(O_\Gamma^+(\mu)) = O_\Upsilon^-(h(\mu)). \tag{13.14}$$

Proof: For an arbitrary $\mu \in \mathbf{B}^n$ we have, by making use of Theorem 3.6, page 42, the following proofs of (13.7) and (13.11):

$$g(h(\mu_\Gamma^-)) = g(h(\mu)_\Phi^-) = g(h(\mu)_\Psi^+) = g(h(\mu))_\Upsilon^+,$$
$$h(\mu_\Gamma^-) = h(\mu_\Phi^+) = h(\mu)_\Psi^+ = h(\mu)_\Upsilon^-.$$

In proving (13.9), (13.13) we use Theorem 13.3, page 149 also and we get:

$$g(h(O_\Gamma^-(\mu))) = g(O_\Phi^-(h(\mu))) = g(O_\Psi^+(h(\mu))) = O_\Upsilon^+(g(h(\mu))),$$
$$h(O_\Gamma^-(\mu)) = h(O_\Phi^+(\mu)) = O_\Psi^+(h(\mu)) = O_\Upsilon^-(h(\mu))$$ □

Remark 13.5 Other combinations of isomorphisms, antiisomorphisms and symmetry are also possible, when we use Theorem 4.4, page 52. For example, if in Theorem 13.5 (a), $(h, h') \in Iso(\Gamma, \Phi)$ is replaced by $(h, h')^{\frown} \in Iso^{\frown}(\Gamma, \Phi)$, then in its proof $h(\mu_\Gamma^-) = h(\mu)_\Phi^-$ is replaced by $h(\mu_\Gamma^-) = h(\mu)_\Phi^+$ etc.

Remark 13.6 If in Theorem 13.5 (a) $g \circ h = 1_{\mathbf{B}^n}$, then the time-reversal symmetry of Γ and Υ follows.

13.5 Other Properties

Remark 13.7 The time-reversal symmetry of $\Phi, \Psi : \mathbf{B}^n \to \mathbf{B}^n$ does not imply the satisfaction of

$$\forall \lambda \in \mathbf{B}^n, \exists \nu \in \mathbf{B}^n, (\Phi^\lambda \circ \Psi^\nu)(\mu) = \mu, \tag{13.15}$$

or

$$\forall \nu \in \mathbf{B}^n, \exists \lambda \in \mathbf{B}^n, (\Psi^\nu \circ \Phi^\lambda)(\mu) = \mu, \tag{13.16}$$

to be compared with (13.3) and (13.4). For this, we notice that in Figure 13.1 with Φ at (a) and Ψ at (b) we have $(0,1)_\Psi^+ = \{(0,0),(0,1)\}$ and for $\lambda = (1,1)$ we get

$$\Phi^\lambda((0,1)_\Psi^+) = \Phi(\{(0,0),(0,1)\}) = \{(1,1)\} \neq \{(0,1)\},$$

in other words (13.15) is false. The theorem that follows shows that (13.15), (13.16) take place under a weaker form.

Theorem 13.6 For $\Phi, \Psi : \mathbf{B}^n \to \mathbf{B}^n$ and $\mu \in \mathbf{B}^n$,

$$\mu_\Phi^- = \mu_\Psi^+, \tag{13.17}$$

implies

$$\forall \lambda \in \mathbf{B}^n, (\Phi^\lambda)^{-1}(\mu) \neq \emptyset \Longrightarrow \exists \nu \in \mathbf{B}^n, (\Phi^\lambda \circ \Psi^\nu)(\mu) = \mu, \tag{13.18}$$

and

$$\mu_{\Psi}^{-} = \mu_{\Phi}^{+}, \tag{13.19}$$

implies

$$\forall \nu \in \mathbf{B}^n, (\Psi^{\nu})^{-1}(\mu) \neq \emptyset \Longrightarrow \exists \lambda \in \mathbf{B}^n, (\Psi^{\nu} \circ \Phi^{\lambda})(\mu) = \mu. \tag{13.20}$$

Proof: (13.17)$\Longrightarrow$(13.18). For $\lambda \in \mathbf{B}^n$ arbitrary, we suppose that $(\Phi^{\lambda})^{-1}(\mu) \neq \emptyset$ and let $\mu' \in (\Phi^{\lambda})^{-1}(\mu)$ be arbitrary too. As $\mu' \in \mu_{\Phi}^{-} = \mu_{\Psi}^{+}$, some $\nu \in \mathbf{B}^n$ exists with $\mu' = \Psi^{\nu}(\mu)$. We have $(\Phi^{\lambda} \circ \Psi^{\nu})(\mu) = \Phi^{\lambda}(\mu') = \mu$, thus (13.18) is true.

(13.19)$\Longrightarrow$(13.20). Similar □

Example 13.4 We continue the example from Remark 13.7. Indeed, for $\lambda = (1,1)$ we have in Figure 13.1 that $\Phi^{-1}(0,1) = \emptyset$. We can take however $\lambda = (0,1)$ giving $(\Phi^{(0,1)})^{-1}(0,1) = \{(0,0),(0,1)\}$ and then for $\nu = (1,1)$ we have

$$(\Phi^{\lambda} \circ \Psi^{\nu})(0,1) = \Phi^{(0,1)}(\Psi(0,1)) = \Phi^{(0,1)}(0,0) = (0,1).$$

14

Time-Reversal Symmetry vs tcpo

We show first that if Φ fulfills tcpo and Φ, Ψ are time-reversed symmetrical, then Ψ fulfills tcpo. We define then, given a function Φ that fulfills the strong tcpo, the unique function Ψ that is the time-reversed symmetrical of Φ and we prove that it fulfills the strong tcpo also.

An interesting result is that if Φ fulfills tcpo, then the necessary and the sufficient condition that a function Ψ, which is time-reversed symmetrical with Φ exists is that Φ fulfills the strong tcpo.

Examples of functions Φ, Ψ that are time-reversed symmetrical and fulfill the strong tcpo are given.

14.1 Time-Reversal Symmetry vs tcpo

Theorem 14.1 If Φ fulfills tcpo and the time-reversal symmetry of Φ, Ψ holds, then Ψ fulfills tcpo.

Proof: We suppose against all reason that Ψ does not satisfy tcpo. Some μ, $p \geq 2$ and $i_1, \ldots, i_p \in \{1, \ldots, n\}$ distinct exist then such that $\Psi(\mu) = \mu \oplus \varepsilon^{i_1} \oplus \ldots \oplus \varepsilon^{i_p}$. Since $\mu \oplus \varepsilon^{i_1} \oplus \ldots \oplus \varepsilon^{i_p} \in \mu_\Psi^+ = \mu_\Phi^-$, some $\lambda \in \mathbf{B}^n$ exists such that $\Phi^\lambda(\mu \oplus \varepsilon^{i_1} \oplus \ldots \oplus \varepsilon^{i_p}) = \mu$ and, with the notation $\mu' = \mu \oplus \varepsilon^{i_1} \oplus \ldots \oplus \varepsilon^{i_p}$, we get $\Phi^\lambda(\mu') = \mu' \oplus \varepsilon^{i_1} \oplus \ldots \oplus \varepsilon^{i_p}$. We have obtained the existence of the set I such that $\{i_1, \ldots, i_p\} \subset I \subset \{1, \ldots, n\}$ and $\Phi(\mu') = \mu' \oplus \underset{i \in I}{\Xi} \varepsilon^i$, where $card(I) \geq 2$. This last assertion represents a contradiction with the request that Φ fulfills tcpo $\qquad\qquad\square$

Remark 14.1 Functions Φ exist that fulfill tcpo and the time-reversal symmetry of Φ and Ψ holds for no function Ψ, see Figure 14.1. Indeed, trying to construct Ψ by reversing the arrows, we face the situation that $2^k - 1, k \in \mathbf{N}$ arrows may start from $(1, 1) : 0, 1, 3, \ldots$ arrows $\neq 2$ arrows of Φ pointing to $(1, 1)$. A consequence of this remark is that, in order to assure the existence of

Boolean Functions: Topics in Asynchronicity, First Edition. Serban E. Vlad.
© 2019 John Wiley & Sons, Inc. Published 2019 by John Wiley & Sons, Inc.

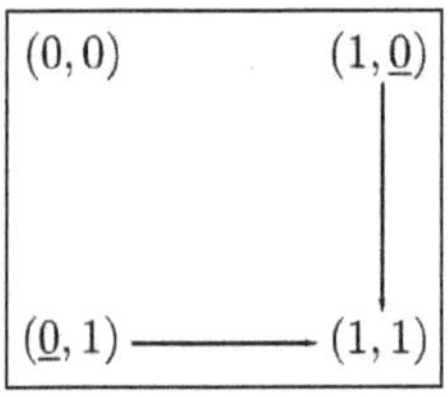

Figure 14.1 Function Φ that fulfills tcpo; the time-reversal symmetry of Φ and Ψ holds for no function Ψ.

time-reversal symmetry under tcpo, stronger requests than the fulfillment of tcpo are necessary.

14.2 Time-Reversal Symmetry vs the Strong tcpo

Theorem 14.2 We suppose that Φ fulfills the strong tcpo and we define $\Psi : \mathbf{B}^n \to \mathbf{B}^n$ by $\forall \mu \in \mathbf{B}^n$,

$$\Psi(\mu) = \begin{cases} \mu, \text{ if } \Phi^{-1}(\mu) = \emptyset, \\ \mu, \text{ if } \Phi^{-1}(\mu) = \{\mu\}, \\ \mu \oplus \varepsilon^i, \text{ if } \exists i \in \{1, \ldots, n\}, \Phi^{-1}(\mu) = \{\mu \oplus \varepsilon^i\}, \\ \mu \oplus \varepsilon^i, \text{ if } \exists i \in \{1, \ldots, n\}, \Phi^{-1}(\mu) = \{\mu, \mu \oplus \varepsilon^i\}. \end{cases} \tag{14.1}$$

Then

(a) the time-reversal symmetry of Φ and Ψ holds;

(b) Ψ fulfills the strong tcpo.

Proof: (a) We notice first of all that Ψ is defined on all of $\mathbf{B}^n$ from Theorem 10.1 (b), page 107 and Definition 10.1, page 108 and that it fulfills tcpo. We fix an arbitrary $\mu \in \mathbf{B}^n$ and we have four possibilities.

Case (i) $\Phi^{-1}(\mu) = \emptyset$, μ is a source for Φ (see Corollary 10.2 (i), page 111), when $\Psi(\mu) = \mu$,

$$\mu_\Phi^- = \{\mu\} = \mu_\Psi^+,$$
$$\exists i \in \{1, \ldots, n\}, \mu_\Phi^+ = \{\mu, \mu \oplus \varepsilon^i\}$$

and we prove $\mu_\Psi^- = \{\mu, \mu \oplus \varepsilon^i\}$.

$\{\mu, \mu \oplus \varepsilon^i\} \subset \mu_\Psi^-$. In this case, $\Phi(\mu) = \mu \oplus \varepsilon^i$ and, because $\mu \in \Phi^{-1}(\mu \oplus \varepsilon^i)$, we have that $\Psi(\mu \oplus \varepsilon^i) = \mu$. It has resulted that $\mu, \mu \oplus \varepsilon^i \in \mu_\Psi^-$.

$\mu_\Psi^- \subset \{\mu, \mu \oplus \varepsilon^i\}$. We suppose against all reason that $\mu' \in \mu_\Psi^-$ exists, $\mu' \neq \mu, \mu' \neq \mu \oplus \varepsilon^i$, in other words, $\mu'' \in \mathbf{B}^n$ and $\nu \in \mathbf{B}^n$ exist such that $\Psi(\mu') = \mu''$ and $\Psi^\nu(\mu') = \mu$. The situation $\mu' = \mu''$ is impossible, as far as it implies the contradiction $\mu' = \Psi(\mu') = \Psi^\nu(\mu') = \mu$, thus $\mu' \neq \mu''$. From the definition of Ψ, see (14.1), we get the existence of $j \in \{1, \ldots, n\}$ with $\mu'' = \mu' \oplus \varepsilon^j$, resulting further that either $\Phi^{-1}(\mu') = \{\mu' \oplus \varepsilon^j\}$ or $\Phi^{-1}(\mu') = \{\mu', \mu' \oplus \varepsilon^j\}$ holds. We have $\{\Psi^\nu(\mu') | \nu \in \mathbf{B}^n\} = \{\mu', \mu' \oplus \varepsilon^j\}$ and $\mu \in \{\mu', \mu' \oplus \varepsilon^j\}$.

But $\mu = \mu'$ is impossible, from the way that we have chosen μ' and the only possibility becomes $\mu = \mu' \oplus \varepsilon^j$.

Case $\Phi^{-1}(\mu') = \{\mu' \oplus \varepsilon^j\}$, i.e. $\Phi^{-1}(\mu \oplus \varepsilon^j) = \{\mu\}$

We infer $\Phi(\mu) = \mu \oplus \varepsilon^j = \mu \oplus \varepsilon^i, j = i$ and $\mu' = \mu \oplus \varepsilon^i$, contradiction.

Case $\Phi^{-1}(\mu') = \{\mu', \mu' \oplus \varepsilon^j\}$, i.e. $\Phi^{-1}(\mu \oplus \varepsilon^j) = \{\mu \oplus \varepsilon^j, \mu\}$

Once again the fact that $\Phi(\mu) = \mu \oplus \varepsilon^j = \mu \oplus \varepsilon^i$ implies $j = i$ and $\mu' = \mu \oplus \varepsilon^i$, representing a contradiction.

It has resulted that such a μ' does not exist.

Case (ii) $\Phi^{-1}(\mu) = \{\mu\}$, μ is an isolated fixed point of Φ (see Corollary 10.2 (ii), page 111), when $\Psi(\mu) = \mu$,

$$\mu_\Phi^- = \{\mu\} = \mu_\Psi^+,$$
$$\mu_\Phi^+ = \{\mu\}$$

and we prove $\mu_\Psi^- = \{\mu\}$.

$\{\mu\} \subset \mu_\Psi^-$. Obvious.

$\mu_\Psi^- \subset \{\mu\}$. We suppose against all reason that the inclusion does not take place and let $\mu' \neq \mu$ with the property that $\mu'' \in \mathbf{B}^n$ and $\nu \in \mathbf{B}^n$ exist with $\Psi(\mu') = \mu''$ and $\Psi^\nu(\mu') = \mu$.

The situation $\mu' = \mu''$ gives the contradiction $\mu' = \Psi(\mu') = \Psi^\nu(\mu') = \mu$, therefore $\mu' \neq \mu''$.

From the way that Ψ was defined in (14.1) some $j \in \{1, \dots, n\}$ exists with $\mu'' = \mu' \oplus \varepsilon^j$ and we have that either $\Phi^{-1}(\mu') = \{\mu' \oplus \varepsilon^j\}$ or $\Phi^{-1}(\mu') = \{\mu', \mu' \oplus \varepsilon^j\}$ is true. We infer $\{\Psi^\nu(\mu') | \nu \in \mathbf{B}^n\} = \{\mu', \mu' \oplus \varepsilon^j\}$ and $\mu \in \{\mu', \mu' \oplus \varepsilon^j\}$, where the only possibility is $\mu = \mu' \oplus \varepsilon^j$.

Case $\Phi^{-1}(\mu') = \{\mu' \oplus \varepsilon^j\}$, i.e. $\Phi^{-1}(\mu \oplus \varepsilon^j) = \{\mu\}$

This implies the contradiction $\Phi(\mu) = \mu \oplus \varepsilon^j$.

Case $\Phi^{-1}(\mu') = \{\mu', \mu' \oplus \varepsilon^j\}$, i.e. $\Phi^{-1}(\mu \oplus \varepsilon^j) = \{\mu \oplus \varepsilon^j, \mu\}$

Once again $\Phi(\mu) = \mu \oplus \varepsilon^j$ represents a contradiction.

The conclusion is that such a μ' does not exist.

Case (iii) some $i \in \{1, \dots, n\}$ exists such that $\Phi^{-1}(\mu) = \{\mu \oplus \varepsilon^i\}$, μ is a transient point of Φ (see Corollary 10.2 (iii), page 111), $\Psi(\mu) = \mu \oplus \varepsilon^i$,

$$\mu_\Phi^- = \{\mu, \mu \oplus \varepsilon^i\} = \mu_\Psi^+,$$
$$\exists j \in \{1, \dots, n\}, \mu_\Phi^+ = \{\mu, \mu \oplus \varepsilon^j\}$$

and we show that $\mu_\Psi^- = \{\mu, \mu \oplus \varepsilon^j\}$.

$\{\mu, \mu \oplus \varepsilon^j\} \subset \mu_\Psi^-$. Since $\Phi(\mu) = \mu \oplus \varepsilon^j$, we get $\Psi(\mu \oplus \varepsilon^j) = \mu$ thus $\mu, \mu \oplus \varepsilon^j \in \mu_\Psi^-$.

$\mu_\Psi^- \subset \{\mu, \mu \oplus \varepsilon^j\}$. Let us suppose against all reason that $\mu' \in \mu_\Psi^-$ exists, having the property that $\mu' \neq \mu$ and $\mu' \neq \mu \oplus \varepsilon^j$. Then $\mu'' \in \mathbf{B}^n$ and $\nu \in \mathbf{B}^n$ exist with $\Psi(\mu') = \mu''$ and $\Psi^\nu(\mu') = \mu$. Like before, the situation $\mu' = \mu''$ is impossible, because it implies $\mu' = \Psi(\mu') = \Psi^\nu(\mu') = \mu$, therefore $\mu' \neq \mu''$.

From the definition (14.1) of Ψ we get the existence of $k \in \{1, \dots, n\}$ with $\mu'' = \mu' \oplus \varepsilon^k$, resulting furthermore that one of $\Phi^{-1}(\mu') = \{\mu' \oplus \varepsilon^k\}$,

$\Phi^{-1}(\mu') = \{\mu', \mu' \oplus \varepsilon^k\}$ is true. On the other hand, $\{\Psi^\nu(\mu') | \nu \in \mathbf{B}^n\} = \{\mu', \mu' \oplus \varepsilon^k\}$ and $\mu \in \{\mu', \mu' \oplus \varepsilon^k\}$. But $\mu = \mu'$ is impossible, thus $\mu = \mu' \oplus \varepsilon^k$.

Case $\Phi^{-1}(\mu') = \{\mu' \oplus \varepsilon^k\}$, i.e. $\Phi^{-1}(\mu \oplus \varepsilon^k) = \{\mu\}$

In this case, $\Phi(\mu) = \mu \oplus \varepsilon^k = \mu \oplus \varepsilon^j, k = j$ and $\mu = \mu' \oplus \varepsilon^j$, contradiction.

Case $\Phi^{-1}(\mu') = \{\mu', \mu' \oplus \varepsilon^k\}$, i.e. $\Phi^{-1}(\mu \oplus \varepsilon^k) = \{\mu \oplus \varepsilon^k, \mu\}$

The fact that $\Phi(\mu) = \mu \oplus \varepsilon^k = \mu \oplus \varepsilon^j$ implies $k = j$ and $\mu = \mu' \oplus \varepsilon^j$, contradiction.

We have obtained that such a μ' does not exist.

Case (iv) some $i \in \{1, \ldots, n\}$ exists such that $\Phi^{-1}(\mu) = \{\mu, \mu \oplus \varepsilon^i\}$, μ is a sink for Φ (see Corollary 10.2 (iv), page 111), $\Psi(\mu) = \mu \oplus \varepsilon^i$,

$$\mu_\Phi^- = \{\mu, \mu \oplus \varepsilon^i\} = \mu_\Psi^+,$$
$$\mu_\Phi^+ = \{\mu\}$$

and we prove that $\mu_\Psi^- = \{\mu\}$.

$\{\mu\} \subset \mu_\Psi^-$. Obvious.

$\mu_\Psi^- \subset \{\mu\}$. We suppose against all reason that $\mu' \in \mu_\Psi^-$ exists, $\mu' \neq \mu$, in other words $\mu'' \in \mathbf{B}^n$ and $\nu \in \mathbf{B}^n$ exist such that $\Psi(\mu') = \mu''$ and $\Psi^\nu(\mu') = \mu$. The situation $\mu' = \mu''$ cannot take place, because it implies the contradiction $\mu' = \Psi(\mu') = \Psi^\nu(\mu') = \mu$. As $\mu' \neq \mu''$ and taking into account (14.1), we obtain the existence of $j \in \{1, \ldots, n\}$ with $\mu'' = \mu' \oplus \varepsilon^j$. This fact shows us furthermore that one of $\Phi^{-1}(\mu') = \{\mu' \oplus \varepsilon^j\}$, $\Phi^{-1}(\mu') = \{\mu', \mu' \oplus \varepsilon^j\}$ is true. As $\{\Psi^\nu(\mu') | \nu \in \mathbf{B}^n\} = \{\mu', \mu' \oplus \varepsilon^j\}$ thus we have $\mu \in \{\mu', \mu' \oplus \varepsilon^j\}$, and because the hypothesis has excluded the case $\mu = \mu'$, we infer that $\mu = \mu' \oplus \varepsilon^j$ is the only possibility.

Case $\Phi^{-1}(\mu') = \{\mu' \oplus \varepsilon^j\}$, i.e. $\Phi^{-1}(\mu \oplus \varepsilon^j) = \{\mu\}$

The contradiction $\mu \oplus \varepsilon^j = \Phi(\mu) = \mu$ results.

Case $\Phi^{-1}(\mu') = \{\mu', \mu' \oplus \varepsilon^j\}$, i.e. $\Phi^{-1}(\mu \oplus \varepsilon^j) = \{\mu \oplus \varepsilon^j, \mu\}$

We infer $\mu \oplus \varepsilon^j = \Phi(\mu) = \mu$, representing a contradiction.

The conclusion is that such a μ' does not exist.

In all the Cases (i)–(iv) we have proved that $\mu_\Phi^- = \mu_\Psi^+, \mu_\Phi^+ = \mu_\Psi^-$ hold, thus the time-reversal symmetry of Φ and Ψ results.

(b) Ψ satisfies tcpo and let $\mu \in \mathbf{B}^n$, $i, j \in \{1, \ldots, n\}$ arbitrary with $\Psi(\mu \oplus \varepsilon^i) = \Psi(\mu \oplus \varepsilon^j) = \mu$ wherefrom, see (14.1), $\mu \in \Phi^{-1}(\mu \oplus \varepsilon^i) \cap \Phi^{-1}(\mu \oplus \varepsilon^j)$. We infer $\Phi(\mu) = \mu \oplus \varepsilon^i = \mu \oplus \varepsilon^j$, thus $i = j$, i.e. Ψ fulfills the strong tcpo □

Remark 14.2 In the previous theorem, Φ, Ψ are time-reversed symmetrical and they satisfy the strong tcpo both. Here are some additional remarks:

- *At Case (i)*: $\Psi(\mu) = \mu$, $\Psi(\mu \oplus \varepsilon^i) = \mu$, $\mu_\Psi^- = \{\mu, \mu \oplus \varepsilon^i\}$ imply $\Psi^{-1}(\mu) = \{\mu, \mu \oplus \varepsilon^i\}$; the sources of Φ are sinks of Ψ,
- *At Case (ii)*: $\Psi(\mu) = \mu$, $\mu_\Psi^- = \{\mu\}$ imply $\Psi^{-1}(\mu) = \{\mu\}$; the isolated fixed points of Φ and Ψ coincide,

• *At Case (iii)*: $\Psi(\mu) = \mu \oplus \varepsilon^i$, $\mu^-_\Psi = \{\mu, \mu \oplus \varepsilon^j\}$ imply $\Psi(\mu \oplus \varepsilon^j) = \mu$ and $\Psi^{-1}(\mu) = \{\mu \oplus \varepsilon^j\}$; the transient points of Φ and Ψ coincide,

• *At Case (iv)*: $\Psi(\mu) = \mu \oplus \varepsilon^i$, $\mu^-_\Psi = \{\mu\}$ imply $\Psi^{-1}(\mu) = \emptyset$; the sinks of Φ are sources of Ψ.

Theorem 14.3 The following properties are equivalent:
(a) Φ fulfills tcpo and Ψ exists such that the time-reversal symmetry of Φ and Ψ is true;
(b) Φ fulfills the strong tcpo.

Proof: (a) $\Longrightarrow$ (b) We suppose against all reason that Φ does not fulfill the strong tcpo. This means the existence of $\mu \in \mathbf{B}^n$ and $i, j \in \{1, \ldots, n\}$ distinct with $\mu \oplus \varepsilon^i, \mu \oplus \varepsilon^j \in \Phi^{-1}(\mu)$, see Theorem 10.1 (a), page 107 and Definition 10.1, page 108. We denote with Ψ the unique function such that the time-reversal symmetry of Φ and Ψ holds (Theorem 13.4, page 151) and we have $\mu \oplus \varepsilon^i, \mu \oplus \varepsilon^j, \mu \oplus \varepsilon^i \oplus \varepsilon^j \in \mu^-_\Phi = \mu^+_\Psi$ (this is the structure of affine space of μ^+_Ψ, see Theorem 2.11, page 31). We have obtained that $\Psi(\mu) = \mu \oplus \underset{k \in I}{\Xi} \varepsilon^k$, where $I \subset \{1, \ldots, n\}$ and $i, j \in I$. Ψ does not fulfill tcpo, contradiction with Theorem 14.1, page 155.
(b) $\Longrightarrow$ (a) We get that Φ fulfills tcpo, and Theorem 14.2 shows how Ψ can be defined such that the time-reversal symmetry of Φ and Ψ holds □

Remark 14.3 Theorem 14.3 implies the following result. Let us suppose that Φ fulfills tcpo. The necessary and the sufficient condition that the time-reversed symmetrical function Ψ of Φ exists is that Φ fulfills the strong tcpo.
This statement shows the necessity and the sufficiency of defining the strong tcpo in order to analyze the time-reversal symmetry under tcpo.

14.3 Examples

Example 14.1 In Figure 9.1, page 100 we have the function $\Phi(\mu_1, \mu_2) = (\mu_1 \cup \mu_2, \overline{\mu_1} \cup \mu_2)$ that fulfills the strong tcpo: $\forall \mu \in \mathbf{B}^n$,

$$\Phi^{-1}(0,0) = \emptyset, \text{see } (10.2)_{page\ 107},$$
$$\Phi^{-1}(0,1) = \{(0,0)\}, \text{see } (10.4)_{page\ 107},$$
$$\Phi^{-1}(1,1) = \{(0,1),(1,1)\}, \text{see } (10.5)_{page\ 107},$$
$$\Phi^{-1}(1,0) = \{(1,0)\}, \text{see } (10.3)_{page\ 107}.$$

The function $\Psi(\mu_1, \mu_2) = (\mu_1 \overline{\mu_2}, \mu_1 \mu_2)$ fulfills:

$$\Psi^{-1}(0,0) = \{(0,0),(0,1)\}, \text{see } (10.5)_{page\ 107},$$

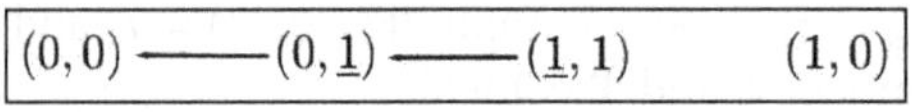

Figure 14.2 The function $\Psi(\mu_1, \mu_2) = (\mu_1 \overline{\mu_2}, \mu_1 \mu_2)$ which is the time-reversed symmetrical of the function Φ from Figure 9.1, page 100.

$$\Psi^{-1}(0,1) = \{(1,1)\}, \text{see } (10.4)_{page\ 107},$$
$$\Psi^{-1}(1,0) = \{(1,0)\}, \text{see } (10.3)_{page\ 107},$$
$$\Psi^{-1}(1,1) = \emptyset, \text{see } (10.2)_{page\ 107}$$

and its state portrait is drawn in Figure 14.2. The time-reversal symmetry of Φ and Ψ holds since, for example:

$$(0,0)^+_\Phi = (0,0)^-_\Psi = \{(0,0), (0,1)\},$$
$$(0,1)^+_\Phi = (0,1)^-_\Psi = \{(0,1), (1,1)\},$$
$$(1,1)^+_\Phi = (1,1)^-_\Psi = \{(1,1)\},$$
$$(1,0)^+_\Phi = (1,0)^-_\Psi = \{(1,0)\}.$$

Example 14.2 The functions $\Phi(\mu_1, \mu_2) = (\mu_1, \overline{\mu_1}\ \overline{\mu_2})$, $\Psi(\mu_1, \mu_2) = (\mu_1, \mu_1 \cup \overline{\mu_2})$ from Figures 14.3 and 14.4 fulfill the strong tcpo, for example $\Phi^{-1}(0,0) = \{(0,1)\}$, $\Phi^{-1}(0,1) = \{(0,0)\}$, $\Phi^{-1}(1,1) = \emptyset$ and $\Phi^{-1}(1,0) = \{(1,0), (1,1)\}$. Φ and Ψ are time-reversed symmetrical.

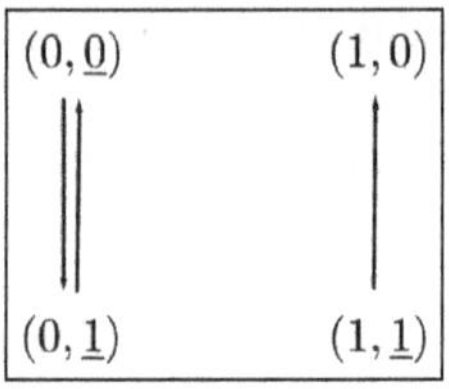

Figure 14.3 The function $\Phi(\mu_1, \mu_2) = (\mu_1, \overline{\mu_1}\ \overline{\mu_2})$.

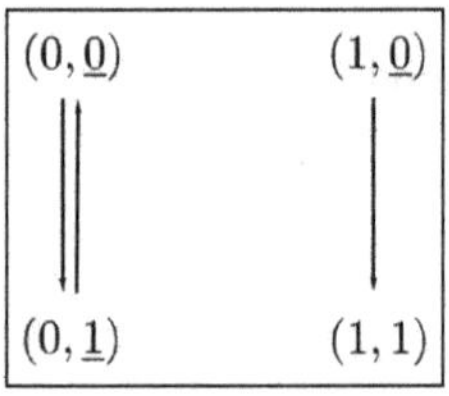

Figure 14.4 The function $\Psi(\mu_1, \mu_2) = (\mu_1, \mu_1 \cup \overline{\mu_2})$.

Example 14.3 We consider the functions $\Phi, \Psi, \Gamma, \Upsilon : \mathbf{B}^2 \to \mathbf{B}^2$ defined in the following way: $\forall \mu \in \mathbf{B}^2$,

$$\Phi(\mu_1, \mu_2) = (\mu_1 \cup \overline{\mu_2}, \mu_2),$$
$$\Psi(\mu_1, \mu_2) = (\mu_1 \cup \overline{\mu_2}, \mu_1 \cup \mu_2),$$

$$\Gamma(\mu_1, \mu_2) = (\mu_1\mu_2, \mu_2),$$
$$\Upsilon(\mu_1, \mu_2) = (\mu_1\mu_2, \overline{\mu_1}\mu_2),$$

whose state portraits were drawn in Figure 14.5 (a), (b) and, respectively, in Figure 14.6 (a), (b). We have the existence of $h, h' : \mathbf{B}^2 \to \mathbf{B}^2$ defined by: $\forall \mu \in \mathbf{B}^2$,

$$h(\mu_1, \mu_2) = (\mu_1, \mu_2),$$
$$h'(\mu_1, \mu_2) = (\mu_1, 0)$$

which satisfy $(h, h') \in Hom(\Phi, \Psi) : \forall \mu \in \mathbf{B}^2, \forall \nu \in \mathbf{B}^2$,

$$\Phi^{(\nu_1,\nu_2)}(\mu_1, \mu_2) = \Psi^{(\nu_1,0)}(\mu_1, \mu_2) = (\overline{\nu_1}\mu_1 \cup \nu_1(\mu_1 \cup \overline{\mu_2}), \mu_2)$$

and $(h, h') \in Hom(\Gamma, \Upsilon) : \forall \mu \in \mathbf{B}^2, \forall \nu \in \mathbf{B}^2$,

$$\Gamma^{(\nu_1,\nu_2)}(\mu_1, \mu_2) = \Upsilon^{(\nu_1,0)}(\mu_1, \mu_2) = (\overline{\nu_1}\mu_1 \cup \nu_1\mu_1\mu_2, \mu_2).$$

On the other hand, Φ, Γ are time-reversed symmetrical and Ψ, Υ are time-reversed symmetrical also and all the four functions fulfill the strong tcpo. The following "diagram" is "commutative":

$$\begin{array}{ccc} \Phi & \overset{time-reversal\ symmetry}{\longleftrightarrow} & \Gamma \\ (h,h') \downarrow & & \downarrow (h,h') \\ \Psi & \overset{time-reversal\ symmetry}{\longleftrightarrow} & \Upsilon \end{array}$$

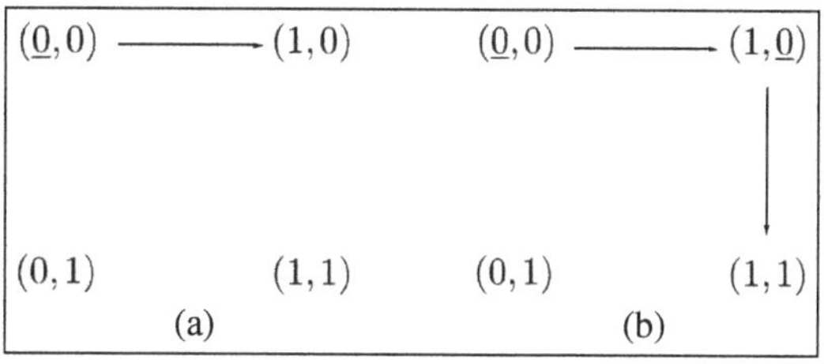

Figure 14.5 $\Phi(\mu_1, \mu_2) = (\mu_1 \cup \overline{\mu_2}, \mu_2)$ at (a) and $\Psi(\mu_1, \mu_2) = (\mu_1 \cup \overline{\mu_2}, \mu_1 \cup \mu_2)$ at (b)

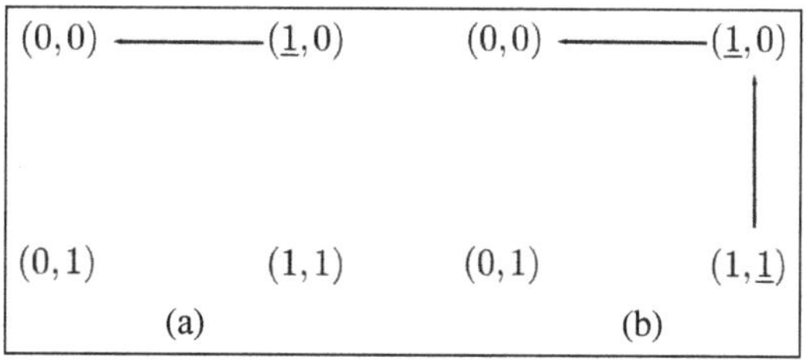

Figure 14.6 $\Gamma(\mu_1, \mu_2) = (\mu_1\mu_2, \mu_2)$ at (a) and $\Upsilon(\mu_1, \mu_2) = (\mu_1\mu_2, \overline{\mu_1}\mu_2)$ at (b).

15

Time-Reversal Symmetry vs the Generalized tcpo

If Φ satisfies the generalized technical condition of proper operation (tcpo), then its time-reversed symmetrical function Ψ might not exist, but if Φ satisfies the strong generalized tcpo, then Ψ exists and, in addition, it satisfies the strong generalized tcpo too.

15.1 Time-Reversal Symmetry vs the Generalized tcpo

Remark 15.1 We get back again to the example given by the function Φ : $\mathbf{B}^n \longrightarrow \mathbf{B}^n$ from Figure 11.3, page 120. This function satisfies the generalized tcpo, but it does not satisfy the strong generalized tcpo, as we have already noticed. Φ does not have a time-reversed symmetrical function Ψ, because by reversing the sense of the arrows in $(0, 0, 0)$ we would have five arrows, and this is impossible, as far as the number of arrows from one point μ can be $2^k - 1$ only, $k \in \{0, 1, 2, \dots\}$. We infer from here that the generalized tcpo does not guarantee the existence of a time-reversed symmetrical function.

Theorem 15.1 We suppose that $\Phi : \mathbf{B}^n \longrightarrow \mathbf{B}^n$ fulfills the strong generalized tcpo and we define $\Psi : \mathbf{B}^n \longrightarrow \mathbf{B}^n$ by $\forall \mu \in \mathbf{B}^n, \Psi(\mu) = \lambda$, where λ is the unique one with $\mu_\Phi^- = [\lambda, \mu]$, see Theorem 12.7, page 141. Then Ψ is the time-reversed symmetrical function of Φ and it fulfills the strong generalized tcpo.

Proof: We fix an arbitrary $\mu \in \mathbf{B}^n$, thus $\lambda \in \mathbf{B}^n$ is uniquely fixed itself. We see that
$$\mu_\Psi^+ = [\mu, \Psi(\mu)] = [\mu, \lambda] = \mu_\Phi^- \tag{15.1}$$

and Theorem 12.6, page 137 shows the existence of several possibilities.

(j) Case $\Phi^{-1}(\mu) = \varnothing, \Phi(\mu) \neq \mu, \mu_\Phi^+ \neq \{\mu\}$,

(j.1) Case $\mu_\Phi^- = \{\mu\} = [\mu, \mu], \Psi(\mu) = \mu$.

Boolean Functions: Topics in Asynchronicity, First Edition. Serban E. Vlad.
© 2019 John Wiley & Sons, Inc. Published 2019 by John Wiley & Sons, Inc.

We claim the truth of one of

$$\Phi^{-1}(\Phi(\mu)) = [\mu, \Phi(\mu)), \tag{15.2}$$

$$\Phi^{-1}(\Phi(\mu)) = [\mu, \Phi(\mu)]. \tag{15.3}$$

In this case, for any $\nu \in (\mu, \Phi(\mu))$, $\Phi^{-1}(\nu) = \emptyset$ and Theorem 12.6 (i.2) shows that $\nu_{\Phi}^{-} = [\mu, \nu]$, thus $\Psi(\nu) = \mu$. If (15.2) holds, we can apply Theorem 12.6 (iii) and if (15.3) holds, we can apply Theorem 12.6 (iv); in both situations we get $\Phi(\mu)_{\Phi}^{-} = [\mu, \Phi(\mu)]$, hence $\Psi(\Phi(\mu)) = \mu$.

If $\Phi(\Phi(\mu)) \neq \Phi(\mu)$, we prove (15.2). $[\mu, \Phi(\mu)) \subset \Phi^{-1}(\Phi(\mu))$ is a consequence of the generalized tcpo fulfilled by Φ.

$\Phi^{-1}(\Phi(\mu)) \subset [\mu, \Phi(\mu))$. We suppose against all reason that this is not true, i.e. $\nu \notin [\mu, \Phi(\mu))$ exists such that $\Phi(\nu) = \Phi(\mu)$. We have obtained the existence of $\delta \in \mathbf{B}^n$ with $\Phi^{-1}(\Phi(\mu)) = [\delta, \Phi(\mu))$ and $\mu \in (\delta, \Phi(\delta))$. From Theorem 12.6 (i.2), we infer, however, that $\mu_{\Phi}^{-} = [\delta, \mu] \neq \{\mu\}$, contradiction. (15.2) is proved.

If $\Phi(\Phi(\mu)) = \Phi(\mu)$, relation (15.3) is proved similarly.

We have shown that $[\mu, \Phi(\mu)] \subset \Psi^{-1}(\mu)$. We state that

$$\Psi^{-1}(\mu) = [\mu, \Phi(\mu)] \tag{15.4}$$

thus the inclusion $\Psi^{-1}(\mu) \subset [\mu, \Phi(\mu)]$ must be proved. We suppose against all reason the existence of $\nu \notin [\mu, \Phi(\mu)]$ such that $\Psi(\nu) = \mu$, meaning from the definition of Ψ that $\nu_{\Phi}^{-} = [\mu, \nu]$, i.e. $\nu \in [\mu, \Phi(\mu)]$, contradiction. (15.4) is proved.

We show that

$$\mu_{\Psi}^{-} = [\mu, \Phi(\mu)]. \tag{15.5}$$

The inclusion $[\mu, \Phi(\mu)] \subset \mu_{\Psi}^{-}$ is clear from (15.4), we prove $\mu_{\Psi}^{-} \subset [\mu, \Phi(\mu)]$. Let against all reason $\nu \notin [\mu, \Phi(\mu)]$ and $\omega \in \mathbf{B}^n$ with $\Psi^{\omega}(\nu) = \mu$, i.e. $\mu \in [\nu, \Psi(\nu)]$. From the definition of Ψ, we infer $\nu_{\Phi}^{-} = [\Psi(\nu), \nu]$ and then $\delta \in \mathbf{B}^n$ exists with $\Phi^{\delta}(\mu) = \nu$, i.e. $\nu \in [\mu, \Phi(\mu)]$, contradiction. (15.5) is proved, hence

$$\mu_{\Psi}^{-} = [\mu, \Phi(\mu)] = \mu_{\Phi}^{+}.$$

The property

$$\forall \nu \in (\mu, \Psi(\mu)), \Psi^{-1}(\nu) = \emptyset \tag{15.6}$$

is trivially satisfied, as far as $(\mu, \Psi(\mu)) = (\mu, \mu) = \emptyset$.

(j.2) Case $\exists \lambda \in \mathbf{B}^n$ such that

$$\Phi^{-1}(\Phi(\lambda)) = [\lambda, \Phi(\lambda)) \text{ or } \Phi^{-1}(\Phi(\lambda)) = [\lambda, \Phi(\lambda)],$$

$\mu \in (\lambda, \Phi(\lambda))$, $\mu_{\Phi}^{-} = [\lambda, \mu]$. In this case, $\Psi(\mu) = \lambda$ and $\forall \nu \in (\lambda, \Phi(\lambda))$, we have similarly $\Psi(\nu) = \lambda$. If $\Phi^{-1}(\Phi(\lambda)) = [\lambda, \Phi(\lambda))$ we use Theorem 12.6 (iii) and if $\Phi^{-1}(\Phi(\lambda)) = [\lambda, \Phi(\lambda)]$ we use Theorem 12.6 (iv) to infer that $\Phi(\lambda)_{\Phi}^{-} = [\lambda, \Phi(\lambda)]$, hence $\Psi(\Phi(\lambda)) = \lambda$.

We prove that

$$\Psi^{-1}(\mu) = \emptyset \tag{15.7}$$

and let us suppose, against all reason, that this is false, i.e. $\nu \in \mathbf{B}^n$ exists with $\Psi(\nu) = \mu$, i.e. $\nu_{\Phi}^{-} = [\mu, \nu]$, thus $\nu \in [\mu, \Phi(\mu)]$. If $\nu = \Phi(\mu)$, then $\Psi(\nu) = \lambda \neq \mu$, contradiction. We infer that $\nu \in [\mu, \Phi(\mu))$. The generalized tcpo of Φ implies $\Phi(\nu) = \Phi(\mu) = \Phi(\lambda)$, thus $\nu \in (\lambda, \Phi(\lambda))$, wherefrom $\Psi(\nu) = \lambda \neq \mu$, contradiction again. (15.7) is proved.

We show that

$$\mu_{\Psi}^{-} = [\mu, \Phi(\mu)] \tag{15.8}$$

and we prove first the inclusion $[\mu, \Phi(\mu)] \subset \mu_{\Psi}^{-}$. Let $\nu \in [\mu, \Phi(\mu)] = [\mu, \Phi(\lambda)]$ arbitrary and we notice that

$$(\lambda \boxplus \mu) \cap (\mu \boxplus \nu) = \emptyset. \tag{15.9}$$

For this, let, against all reason, $i \in (\lambda \boxplus \mu) \cap (\mu \boxplus \nu)$. As $\lambda \boxplus \mu \subset \lambda \boxplus \Phi(\lambda)$ $(\mu \in [\lambda, \Phi(\lambda)])$, we infer that $\lambda_i \neq \mu_i$ and $\lambda_i \neq \Phi_i(\lambda)$, i.e. $\mu_i = \Phi_i(\lambda)$. But $\mu \boxplus \nu \subset \mu \boxplus \Phi(\lambda)$ $(\nu \in [\mu, \Phi(\lambda)])$, wherefrom $\mu_i \neq \nu_i$ and $\mu_i \neq \Phi_i(\lambda)$, contradiction. (15.9) holds. We have $\Psi(\nu) = \lambda$, thus the equation $\Psi^{\omega}(\nu) = \mu (= \lambda \oplus \underset{i \in \mu \boxplus \lambda}{\Xi} \varepsilon^i)$ with the unknown $\omega \in \mathbf{B}^n$ has the solution $\forall j \in \{1, \dots, n\}, \omega_j = \begin{cases} 1, & \text{if } j \in \mu \boxplus \nu, \\ 0, & \text{otherwise} \end{cases}$ allowing $\Psi_j(\nu)$ be computed for $j \in \mu \boxplus \nu$ and be not computed for $j \in \lambda \boxplus \mu$, possibility given by (15.9). Indeed, we have for any $j \in \{1, \dots, n\}$:

$$\Psi_j^{\omega}(\nu) = \begin{cases} \Psi_j(\nu), & \text{if } \omega_j = 1 \\ \nu_j, & \text{otherwise} \end{cases} = \begin{cases} \Psi_j(\nu), & \text{if } j \in \mu \boxplus \nu \\ \nu_j, & \text{otherwise} \end{cases}$$

$$= \begin{cases} \nu_j \oplus 1, & \text{if } j \in \mu \boxplus \nu \\ \nu_j, & \text{otherwise} \end{cases} = \mu_j.$$

We prove now $\mu_{\Psi}^{-} \subset [\mu, \Phi(\mu)]$. Let against all reason $\nu \notin [\mu, \Phi(\mu)]$ and $\omega \in \mathbf{B}^n$ such that $\Psi^{\omega}(\nu) = \mu$, meaning that $\mu \in [\nu, \Psi(\nu)]$. The definition of Ψ implies $\nu_{\Phi}^{-} = [\Psi(\nu), \nu]$, therefore $\delta \in \mathbf{B}^n$ exists with $\Phi^{\delta}(\mu) = \nu$. We have obtained the contradiction $\nu \in [\mu, \Phi(\mu)]$. (15.8) is proved, thus

$$\mu_{\Psi}^{-} = [\mu, \Phi(\mu)] = \mu_{\Phi}^{+}.$$

We claim that the property (15.6) holds. Indeed, $(\mu, \Psi(\mu)) = (\mu, \lambda)$ and for any $\nu \in (\mu, \lambda)$, we have $\nu \in (\lambda, \Phi(\lambda))$. We prove that

$$\Psi^{-1}(\nu) = \emptyset. \tag{15.10}$$

Let us suppose against all reason that $\omega \in \mathbf{B}^n$ exists with $\Psi(\omega) = \nu$, thus $\omega_{\Phi}^{-} = [\nu, \omega]$, in other words $\omega \in [\nu, \Phi(\nu)]$. If $\omega = \Phi(\nu) (= \Phi(\lambda))$, then $\Psi(\omega) = \Psi(\Phi(\lambda)) = \lambda \neq \nu$, contradiction. We infer from here that $\omega \in [\nu, \Phi(\nu))$. From the generalized tcpo, we obtain $\Phi(\omega) = \Phi(\nu) = \Phi(\lambda)$, i.e. $\omega \in [\nu, \Phi(\lambda)) \subset (\lambda, \Phi(\lambda))$, therefore $\Psi(\omega) = \lambda \neq \nu$, contradiction. Statement (15.10) holds, thus (15.6) is true.

(jj) Case $\Phi^{-1}(\mu) = \{\mu\}, \mu_{\Phi}^{-} = \{\mu\}, \Phi(\mu) = \mu, \mu_{\Phi}^{+} = \{\mu\}, \Psi(\mu) = \mu$.

We show that

$$\Psi^{-1}(\mu) = \{\mu\} \tag{15.11}$$

and let us suppose, against all reason that $\nu \in \mathbf{B}^n$ exists, $\nu \neq \mu$, such that $\Psi(\nu) = \mu$. This has its origin in $\nu_{\Phi}^{-} = [\mu, \nu]$. We have obtained the contradiction $\nu \in (\mu, \Phi(\mu)] = (\mu, \mu] = \emptyset$, showing the truth of (15.11).

We prove that

$$\mu_{\Psi}^{-} = \{\mu\}. \tag{15.12}$$

Since $\{\mu\} \subset \mu_{\Psi}^{-}$ is obvious, we prove $\mu_{\Psi}^{-} \subset \{\mu\}$. Let against all reason $\nu \in \mathbf{B}^n, \nu \neq \mu$ and $\omega \in \mathbf{B}^n$ with $\Psi^{\omega}(\nu) = \mu$. Since $\Psi(\nu) \neq \mu$ (otherwise, $\nu \in \Psi^{-1}(\mu) = \{\mu\}$, contradiction), we get $\mu \in (\nu, \Psi(\nu))$. But the definition of Ψ shows that $\nu_{\Phi}^{-} = [\Psi(\nu), \nu]$, wherefrom we get the existence of $\delta \in \mathbf{B}^n$ with $\Phi^{\delta}(\mu) = \nu$. We have obtained the contradiction $\mu = \Phi(\mu) = \Phi^{\delta}(\mu) = \nu$, showing the truth of (15.12). We infer

$$\mu_{\Psi}^{-} = \{\mu\} = \mu_{\Phi}^{+}.$$

Property (15.6) is trivially fulfilled since $(\mu, \Psi(\mu)) = (\mu, \mu) = \emptyset$.

(jjj) Case $\exists \lambda \in \mathbf{B}^n, \Phi^{-1}(\mu) = [\lambda, \mu)$, where $\lambda \neq \mu, \mu_{\Phi}^{-} = [\lambda, \mu], \Phi(\mu) \neq \mu$ and $\Psi(\mu) = \lambda$.

We show that

$$\Phi^{-1}(\Phi(\mu)) = [\mu, \Phi(\mu)) \text{ or } \Phi^{-1}(\Phi(\mu)) = [\mu, \Phi(\mu)] \tag{15.13}$$

is true. We suppose against all reason the existence of $\nu \in \mathbf{B}^n, \nu \neq \mu$, with

$$\Phi^{-1}(\Phi(\mu)) = [\nu, \Phi(\mu)) \text{ or } \Phi^{-1}(\Phi(\mu)) = [\nu, \Phi(\mu)]$$

and we infer $\nu \notin [\mu, \Phi(\mu)]$. As $\mu \neq \Phi(\mu)$ and $\mu \in \Phi^{-1}(\Phi(\mu))$, we have $\mu \in (\nu, \Phi(\mu)) = (\nu, \Phi(\nu))$, wherefrom $\Phi^{-1}(\mu) = \emptyset$, because Φ fulfills the strong generalized tcpo. This is in contradiction with the hypothesis. (15.13) holds.

Let now $\nu \in (\mu, \Phi(\mu))$ arbitrary. We have $\Phi^{-1}(\nu) = \emptyset$ (from the strong generalized tcpo), thus Theorem 12.6 (i.2) implies $\nu_{\Phi}^{-} = [\mu, \nu]$, wherefrom $\Psi(\nu) = \mu$. If in (15.13) $\Phi^{-1}(\Phi(\mu)) = [\mu, \Phi(\mu))$, we can use Theorem 12.6 (iii) and if in (15.13) $\Phi^{-1}(\Phi(\mu)) = [\mu, \Phi(\mu)]$, we can use Theorem 12.6 (iv); we infer in both situations $\Phi(\mu)_{\Phi}^{-} = [\mu, \Phi(\mu)]$, thus $\Psi(\Phi(\mu)) = \mu$. We have proved that $(\mu, \Phi(\mu)] \subset \Psi^{-1}(\mu)$ and we prove now that

$$\Psi^{-1}(\mu) = [\Phi(\mu), \mu). \tag{15.14}$$

In order to prove the inclusion $\Psi^{-1}(\mu) \subset [\Phi(\mu), \mu)$, we suppose against all reason its falsity; then $\nu \notin [\Phi(\mu), \mu)$ exists such that $\Psi(\nu) = \mu$. We have $\nu_{\Phi}^{-} = [\mu, \nu]$, i.e. $\nu \in [\mu, \Phi(\mu)]$. The only possibility is $\nu = \mu$, but this implies the contradiction $\mu_{\Phi}^{-} = \{\mu\}$. Statement (15.14) is proved.

We have also proved that $[\Phi(\mu), \mu] \subset \mu_{\Psi}^{-}$ and we prove now the equality

$$\mu_{\Psi}^{-} = [\Phi(\mu), \mu]. \tag{15.15}$$

Let us suppose, against all reason, that $\mu_{\Psi}^{-} \subset [\Phi(\mu), \mu]$ is false; then $\nu \notin [\Phi(\mu), \mu]$ and $\omega \in \mathbf{B}^n$ exist with $\Psi^{\omega}(\nu) = \mu$, in other words $\mu \in [\nu, \Psi(\nu)]$. We have $\mu \neq \nu$ from the hypothesis and we have also $\mu \neq \Psi(\nu)$ (otherwise we get $\nu \in \Psi^{-1}(\mu) = (\mu, \Phi(\mu)]$, contradiction), therefore $\mu \in (\nu, \Psi(\nu))$. The definition of Ψ in ν gives $\nu_{\Phi}^{-} = [\Psi(\nu), \nu]$, and as $\mu \in (\Psi(\nu), \nu)$, we infer the existence of $\delta \in \mathbf{B}^n$ with $\Phi^{\delta}(\mu) = \nu$, hence $\nu \in [\mu, \Phi(\mu)]$, contradiction again. Statement (15.15) is true, thus

$$\mu_{\Psi}^{-} = [\Phi(\mu), \mu] = \mu_{\Phi}^{+}.$$

In order to prove the truth of (15.6), we prove first that $\forall \nu \in (\lambda, \mu), \Psi(\nu) = \lambda$ and we fix $\nu \in (\Psi(\mu), \mu) = (\lambda, \mu) = (\lambda, \Phi(\lambda))$ arbitrary. The hypothesis of Theorem 12.6 (i.2) is fulfilled, because $\Phi^{-1}(\nu) = \emptyset$ (from the strong generalized tcpo), $\nu \neq \Phi(\nu) = \mu$ (from the generalized tcpo), and (15.13) takes place under the form $\Phi^{-1}(\Phi(\lambda)) = \Phi^{-1}(\mu) = [\lambda, \mu)$. We infer that $\nu_{\Phi}^{-} = [\lambda, \nu]$, wherefrom $\Psi(\nu) = \lambda$.

We keep $\nu \in (\mu, \Psi(\mu)) = (\mu, \lambda)$ arbitrary, fixed and we show that $\Psi^{-1}(\nu) = \emptyset$ holds. Let us suppose against all reason the existence of $\delta \in \mathbf{B}^n$ such that $\Psi(\delta) = \nu$, wherefrom $\delta_{\Phi}^{-} = [\nu, \delta]$ i.e. $\exists \omega \in \mathbf{B}^n$ with $\Phi^{\omega}(\nu) = \delta$. But in this situation $\delta \in [\nu, \Phi(\nu)] = [\nu, \mu] \subset (\lambda, \mu]$. The possibility $\delta = \mu$ implies $[\lambda, \mu] = \mu_{\Phi}^{-} = [\nu, \mu]$ contradiction, thus $\delta \neq \mu$. We have obtained $\delta \in (\lambda, \mu)$, hence $\Psi(\delta) = \lambda$. The last conclusion is in contradiction with the supposition that $\nu = \Psi(\delta) \in (\mu, \lambda)$. Statement (15.6) is proved.

(jv) Case $\exists \lambda \in \mathbf{B}^n, \Phi^{-1}(\mu) = [\lambda, \mu]$, where $\lambda \neq \mu, \mu_{\Phi}^{-} = [\lambda, \mu], \Phi(\mu) = \mu, \Psi(\mu) = \lambda$.

We prove

$$\Psi^{-1}(\mu) = \emptyset \tag{15.16}$$

and let us suppose that this is not true, i.e. $\nu \in \mathbf{B}^n$ exists with $\Psi(\nu) = \mu$, hence $\nu_{\Phi}^{-} = [\mu, \nu]$, wherefrom $\nu \in [\mu, \Phi(\mu)] = \{\mu\}$. But $\Psi(\mu) = \lambda$, contradiction showing the truth of (15.16).

We prove

$$\mu_{\Psi}^{-} = \{\mu\}. \tag{15.17}$$

As $\{\mu\} \subset \mu_{\Psi}^{-}$ is obvious, we prove $\mu_{\Psi}^{-} \subset \{\mu\}$. If, against all reason, the last inclusion is false, then $\nu \in \mathbf{B}^n, \nu \neq \mu$ and $\omega \in \mathbf{B}^n$ exist with $\Psi^{\omega}(\nu) = \mu$, i.e. $\mu \in (\nu, \Psi(\nu)]$. We have from the definition of Ψ in ν : $\nu_{\Phi}^{-} = [\Psi(\nu), \nu]$, in other words $\exists \delta \in \mathbf{B}^n$ with $\Phi^{\delta}(\mu) = \nu$. But $\mu = \Phi(\mu) = \Phi^{\delta}(\mu) = \nu$ represents a contradiction. (15.17) holds and its truth implies:

$$\mu_{\Psi}^{-} = \{\mu\} = \mu_{\Phi}^{+}.$$

In order to prove the satisfaction of (15.6), we notice that $\forall \nu \in (\lambda, \mu)$, $\nu^-_\Phi = [\lambda, \nu]$, from Theorem 12.6 (i.2), thus $\Psi(\nu) = \lambda$. Let an arbitrary $\nu \in (\mu, \Psi(\mu)) = (\mu, \lambda)$. We suppose against all reason the existence of $\delta \in \mathbf{B}^n$ with $\Psi(\delta) = \nu$, thus $\delta^-_\Phi = [\nu, \delta]$ and $\delta \in [\nu, \Phi(\nu)] = [\nu, \mu] \subset [\mu, \lambda)$. But $\Psi(\delta) = \lambda$, representing a contradiction. (15.6) holds.

We finally refer to the definitions

$$((12.3)_{\text{page 131}} \text{ or} \dots \text{or } (12.6)_{\text{page 131}}) \text{ and } (12.15)_{\text{page 133}},$$

$$(13.1)_{\text{page 148}} \text{ and } (13.2)_{\text{page 148}},$$

with $\mu \in \mathbf{B}^n$ arbitrary, fixed, of the strong generalized tcpo and of the time-reversal symmetry and we see that these properties are true in all the previous cases □

Remark 15.2 If Φ, Ψ are time-reversed symmetrical and if they satisfy the strong generalized tcpo, then:

(a) the sources of Φ are sinks of Ψ, Case (j.1) of the previous proof;

(b) the asynchronous transient points of Φ are asynchronous transient points of Ψ, Case (j.2) of the proof;

(c) the isolated fixed points of Φ and Ψ coincide, Case (jj) of the proof;

(d) the synchronous transient points of Φ and Ψ coincide, Case (jjj) of the proof;

(e) the sinks of Φ are sources of Ψ, Case (jv) of the proof.

Remark 15.3 The following statement: "We suppose that Φ fulfills the generalized tcpo. Then the time-reversed symmetrical function Ψ of Φ exists if and only if Φ fulfills the strong generalized tcpo", which is analogue with Remark 14.3, page 159, could not be proved so far. In this framework, strengthening the generalized tcpo to the strong generalized tcpo gives a condition of sufficiency in order that the time-reversed symmetrical function exists.

15.2 Examples

Example 15.1 The functions Φ, Ψ from Example 13.2 and Figure 13.1 (a), (b), page 150 are time-reversed symmetrical and satisfy the strong generalized tcpo.

Example 15.2 The two functions Φ, Ψ from Example 13.3, Figures 13.2 and 13.3, page 150 are time-reversed symmetrical and satisfy the strong generalized tcpo.

Example 15.3 We define the functions $\Phi, \Psi, \Gamma, \Upsilon : \mathbf{B}^2 \to \mathbf{B}^2, \forall \mu \in \mathbf{B}^2$,

$$\Phi(\mu_1, \mu_2) = (1, \mu_2),$$

$$\Psi(\mu_1, \mu_2) = (1, 1),$$
$$\Gamma(\mu_1, \mu_2) = (0, \mu_2),$$
$$\Upsilon(\mu_1, \mu_2) = (0, 0),$$

see Figure 15.1 for Φ, Ψ and Figure 15.2 for Γ, Υ. Given $h, h' : \mathbf{B}^2 \to \mathbf{B}^2$, $\forall \mu \in \mathbf{B}^2$,

$$h(\mu_1, \mu_2) = (\mu_1, \mu_2),$$
$$h'(\mu_1, \mu_2) = (\mu_1, 0)$$

we see that $\forall \mu \in \mathbf{B}^2, \forall \nu \in \mathbf{B}^2$,

$$\Phi^{(\nu_1,\nu_2)}(\mu_1, \mu_2) = \Psi^{(\nu_1,0)}(\mu_1, \mu_2) = (\overline{\nu_1}\mu_1 \cup \nu_1, \mu_2),$$
$$\Gamma^{(\nu_1,\nu_2)}(\mu_1, \mu_2) = \Upsilon^{(\nu_1,0)}(\mu_1, \mu_2) = (\overline{\nu_1}\mu_1, \mu_2),$$

i.e. $(h, h') \in Hom(\Phi, \Psi)$ and $(h, h') \in Hom(\Gamma, \Upsilon)$ hold. On the other hand, Φ, Γ are time-reversed symmetrical and Ψ, Υ are time-reversed symmetrical also. The following "diagram"

$$\begin{array}{ccc} \Phi & \overset{\textit{time-reversal symmetry}}{\longleftrightarrow} & \Gamma \\ (h,h')\downarrow & & \downarrow(h,h') \\ \Psi & \overset{\textit{time-reversal symmetry}}{\longleftrightarrow} & \Upsilon \end{array}$$

of functions that fulfill the strong generalized tcpo is "commutative".

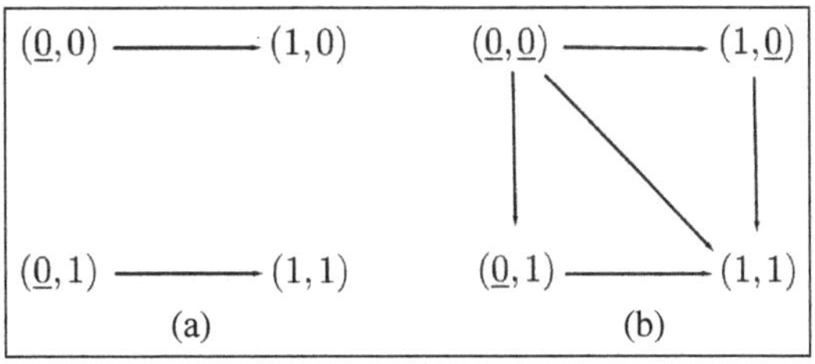

Figure 15.1 $\Phi(\mu_1, \mu_2) = (1, \mu_2)$ at (a) and $\Psi(\mu_1, \mu_2) = (1, 1)$ at (b).

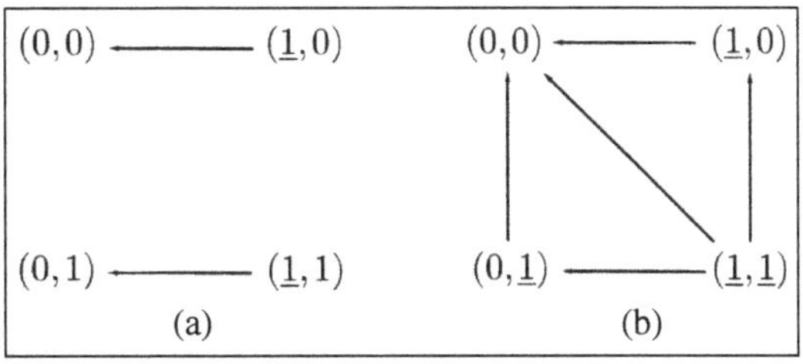

Figure 15.2 $\Gamma(\mu_1, \mu_2) = (0, \mu_2)$ at (a) and $\Upsilon(\mu_1, \mu_2) = (0, 0)$ at (b).

Appendix A

The Category *As*

We define the category[1] *As* whose objects are functions $\Phi : \mathbf{B}^n \longrightarrow \mathbf{B}^n$, and we show that it has finite products.

Definition A.1   The category *As* of the asynchronous systems is defined in the following way:
- the objects of *As* are functions $\Phi : \mathbf{B}^n \longrightarrow \mathbf{B}^n$;
- let the functions $\Phi : \mathbf{B}^n \longrightarrow \mathbf{B}^n$ and $\Psi : \mathbf{B}^m \longrightarrow \mathbf{B}^m$. A morphism from Φ to Ψ is a couple of functions (h, h'), where $h, h' : \mathbf{B}^n \longrightarrow \mathbf{B}^m$ such that $\forall \nu \in \mathbf{B}^n$, the diagram

$$
\begin{array}{ccc}
\mathbf{B}^n & \xrightarrow{\Phi^\nu} & \mathbf{B}^n \\
h \downarrow & & \downarrow h \\
\mathbf{B}^m & \xrightarrow{\Psi^{h'(\nu)}} & \mathbf{B}^m
\end{array}
$$

is commutative. We denote $(h, h') : \Phi \longrightarrow \Psi$. If $\Gamma : \mathbf{B}^p \longrightarrow \mathbf{B}^p$ and $(g, g') : \Psi \longrightarrow \Gamma$ is a morphism, then the composition of the morphisms is defined by

$$(g, g') \circ (h, h') = (g \circ h, g' \circ h').$$

For any $\Phi \in Ob(As)$, the identity morphism $1_\Phi : \Phi \longrightarrow \Phi$ is $1_\Phi = (1_{\mathbf{B}^n}, 1_{\mathbf{B}^n})$.

Definition A.2   We consider the functions $\Phi : \mathbf{B}^n \longrightarrow \mathbf{B}^n, \Psi : \mathbf{B}^m \longrightarrow \mathbf{B}^m$. Their **Cartesian product** is the function $\Phi \times \Psi : \mathbf{B}^{n+m} \longrightarrow \mathbf{B}^{n+m}$ defined in the following way: $\forall \mu \in \mathbf{B}^n, \forall \mu' \in \mathbf{B}^m$,

$$(\Phi \times \Psi)(\mu, \mu') = (\Phi_1(\mu), \dots, \Phi_n(\mu), \Psi_1(\mu'), \dots, \Psi_m(\mu')).$$

We identify $(\mu_1, \dots, \mu_n, \mu'_1, \dots, \mu'_m) \in \mathbf{B}^{n+m}$ and $((\mu_1, \dots, \mu_n), (\mu'_1, \dots, \mu'_m)) \in \mathbf{B}^n \times \mathbf{B}^m$.

1 Categories were introduced in 1945 by Samuel Eilenberg and Saunders Mac Lane.

Boolean Functions: Topics in Asynchronicity, First Edition. Serban E. Vlad.
© 2019 John Wiley & Sons, Inc. Published 2019 by John Wiley & Sons, Inc.

Definition A.3 Let now the functions Φ, Ψ considered as objects $\Phi, \Psi \in Ob(As)$. Their **product** is an object $P \in Ob(As)$ together with two morphisms $(h, h') : P \longrightarrow \Phi, (g, g') : P \longrightarrow \Psi$ such that for any object $\Gamma \in Ob(As)$ and any morphisms $(u, u') : \Gamma \longrightarrow \Phi, (v, v') : \Gamma \longrightarrow \Psi$, there is a unique morphism $(\varphi, \varphi') : \Gamma \longrightarrow P$ such that

$$(h, h') \circ (\varphi, \varphi') = (u, u'), \tag{A.1}$$

$$(g, g') \circ (\varphi, \varphi') = (v, v'). \tag{A.2}$$

The morphisms $(h, h'), (g, g')$ are called the **canonical projections** of P on Φ, Ψ.

Theorem A.1 For any objects $\Phi, \Psi \in Ob(As)$, their product P exists.

Proof: We define $P = \Phi \times \Psi, h, h' : \mathbf{B}^{n+m} \longrightarrow \mathbf{B}^n$ by $\forall \mu \in \mathbf{B}^n, \forall \mu' \in \mathbf{B}^m$,

$$h(\mu, \mu') = h'(\mu, \mu') = \mu \tag{A.3}$$

and also $g, g' : \mathbf{B}^{n+m} \longrightarrow \mathbf{B}^m$ by $\forall \mu \in \mathbf{B}^n, \forall \mu' \in \mathbf{B}^m$,

$$g(\mu, \mu') = g'(\mu, \mu') = \mu'. \tag{A.4}$$

We get that the diagrams

$$\begin{array}{ccc} \mathbf{B}^{n+m} & \overset{(\Phi\times\Psi)^{(\nu,\nu')}}{\longrightarrow} & \mathbf{B}^{n+m} \\ h \downarrow & & \downarrow h \\ \mathbf{B}^n & \overset{\Phi^{h'(\nu,\nu')}}{\longrightarrow} & \mathbf{B}^n \end{array}$$

$$\begin{array}{ccc} \mathbf{B}^{n+m} & \overset{(\Phi\times\Psi)^{(\nu,\nu')}}{\longrightarrow} & \mathbf{B}^{n+m} \\ g \downarrow & & \downarrow g \\ \mathbf{B}^m & \overset{\Psi^{g'(\nu,\nu')}}{\longrightarrow} & \mathbf{B}^m \end{array}$$

commute for all $\nu \in \mathbf{B}^n, \nu' \in \mathbf{B}^m$. In order to see this, we fix $\mu, \nu \in \mathbf{B}^n, \mu', \nu' \in \mathbf{B}^m$ arbitrary and we have

$$\begin{aligned}(\Phi^{h'(\nu,\nu')} \circ h)(\mu, \mu') &= \Phi^{\nu}(h(\mu, \mu')) = \Phi^{\nu}(\mu) \\ &= h(\Phi^{\nu}(\mu), \Psi^{\nu'}(\mu')) = (h \circ (\Phi \times \Psi)^{(\nu,\nu')})(\mu, \mu').\end{aligned}$$

We have proved that $(h, h') : \Phi \times \Psi \longrightarrow \Phi$ is a morphism and the fact that $(g, g') : \Phi \times \Psi \longrightarrow \Psi$ is a morphism also is similarly proved.

Let $\Gamma \in Ob(As)$ arbitrary, $\Gamma : \mathbf{B}^p \longrightarrow \mathbf{B}^p$, together with the morphisms $(u, u') : \Gamma \longrightarrow \Phi, (v, v') : \Gamma \longrightarrow \Psi$ which are arbitrary too. This means that for any $\nu'' \in \mathbf{B}^p$, the diagrams

$$\begin{array}{ccc} \mathbf{B}^p & \overset{\Gamma^{\nu''}}{\longrightarrow} & \mathbf{B}^p \\ u \downarrow & & \downarrow u \\ \mathbf{B}^n & \overset{\Phi^{u'(\nu'')}}{\longrightarrow} & \mathbf{B}^n \end{array}$$

$$\begin{array}{ccc} \mathbf{B}^p & \overset{\Gamma^{\nu''}}{\longrightarrow} & \mathbf{B}^p \\ v \downarrow & & \downarrow v \\ \mathbf{B}^m & \overset{\Psi^{v'(\nu'')}}{\longrightarrow} & \mathbf{B}^m \end{array}$$

are commutative. We define $\varphi, \varphi' : \mathbf{B}^p \longrightarrow \mathbf{B}^{n+m}$ by $\forall \mu'' \in \mathbf{B}^p$,

$$\varphi(\mu'') = (u(\mu''), v(\mu'')), \tag{A.5}$$

$$\varphi'(\mu'') = (u'(\mu''), v'(\mu'')). \tag{A.6}$$

We must show the commutativity of

$$\begin{array}{ccc} \mathbf{B}^p & \overset{\Gamma^{\nu''}}{\longrightarrow} & \mathbf{B}^p \\ \varphi \downarrow & & \downarrow \varphi \\ \mathbf{B}^{n+m} & \overset{(\Phi\times\Psi)^{\varphi'(\nu'')}}{\longrightarrow} & \mathbf{B}^{n+m} \end{array}$$

for all $\nu'' \in \mathbf{B}^p$, i.e. $(\varphi, \varphi') : \Gamma \longrightarrow \Phi \times \Psi$ is a morphism, that (A.1) and (A.2) are true, and that (φ, φ') is unique with these properties. For this, we fix $\mu'', \nu'' \in \mathbf{B}^p$ arbitrary. We have

$$\begin{aligned} (\Phi \times \Psi)^{\varphi'(\nu'')}(\varphi(\mu'')) &= (\Phi \times \Psi)^{(u'(\nu''), v'(\nu''))}(u(\mu''), v(\mu'')) \\ &= (\Phi^{u'(\nu'')}(u(\mu'')), \Psi^{v'(\nu'')}(v(\mu''))) \\ &= (u(\Gamma^{\nu''}(\mu'')), v(\Gamma^{\nu''}(\mu''))) = \varphi(\Gamma^{\nu''}(\mu'')) \end{aligned}$$

and on the other hand

$$\begin{aligned} ((h, h') \circ (\varphi, \varphi'))(\mu'') &= ((h \circ \varphi)(\mu''), (h' \circ \varphi')(\mu'')) \\ &= (h(\varphi(\mu'')), h'(\varphi'(\mu''))) \\ &= (h(u(\mu''), v(\mu'')), h'(u'(\mu''), v'(\mu''))) \\ &= (u(\mu''), u'(\mu'')) = (u, u')(\mu'') \end{aligned}$$

thus (A.1) is true and the proof of (A.2) is similar.

We show the uniqueness of (φ, φ'), and we suppose the existence of another morphism $(\eta, \eta') : \Gamma \longrightarrow \Phi \times \Psi$ with the property that

$$(h, h') \circ (\eta, \eta') = (u, u'), \tag{A.7}$$

$$(g, g') \circ (\eta, \eta') = (v, v') \tag{A.8}$$

are true. For an arbitrary $\mu'' \in \mathbf{B}^p$, we can write

$$\begin{aligned} (u(\mu''), u'(\mu'')) &\overset{(A.7)}{=} ((h, h') \circ (\eta, \eta'))(\mu'') \\ &= ((h \circ \eta)(\mu''), (h' \circ \eta')(\mu'')) \\ &\overset{(A.3)}{=} (\eta_1(\mu''), \dots, \eta_n(\mu''), \eta'_1(\mu''), \dots, \eta'_n(\mu'')), \end{aligned} \tag{A.9}$$

$$\begin{aligned}(v(\mu''), v'(\mu'')) &\overset{(A.8)}{=} ((g, g') \circ (\eta, \eta'))(\mu'') \\ &= ((g \circ \eta)(\mu''), (g' \circ \eta')(\mu'')) \\ &\overset{(A.4)}{=} (\eta_{n+1}(\mu''), \dots, \eta_{n+m}(\mu''), \eta'_{n+1}(\mu''), \dots, \eta'_{n+m}(\mu''))\end{aligned} \tag{A.10}$$

therefore we have:

$$\begin{aligned}\eta(\mu'') &= (\eta_1(\mu''), \dots, \eta_n(\mu''), \eta_{n+1}(\mu''), \dots, \eta_{n+m}(\mu'')) \\ &\overset{(A.9),(A.10)}{=} (u(\mu''), v(\mu'')) \overset{(A.5)}{=} \varphi(\mu''), \\ \eta'(\mu'') &= (\eta'_1(\mu''), \dots, \eta'_n(\mu''), \eta'_{n+1}(\mu''), \dots, \eta'_{n+m}(\mu'')) \\ &\overset{(A.9),(A.10)}{=} (u'(\mu''), v'(\mu'')) \overset{(A.6)}{=} \varphi'(\mu'')\end{aligned}$$

□

Remark A.1 A category is said to be with finite products if any finite family of objects has a product. The previous theorem showed that for finitely many $\Phi, \dots, \Psi \in Ob(As)$, the product $\Phi \times \dots \times \Psi \in Ob(As)$ exists, thus As has finite products. Considering infinitely many $\Phi, \dots, \Psi, \dots \in Ob(As)$ is impossible, since the Cartesian product $\Phi \times \dots \times \Psi \times \dots$ is not an object of As.

Appendix B

Notations

$\mathbf{B} = \{0, 1\}$, the Boolean algebra with two elements

$supp\ a = \{i | i \in I, a_i \neq (0, \dots, 0)\}$, the support set of a

$\underset{i \in I}{\Xi} a_i = \begin{cases} 1, & \text{if } card(supp\ a) \text{ is odd,} \\ 0, & \text{if } card(supp\ a) \text{ is even} \end{cases}$, the modulo 2 summation of a

$\varepsilon^i = (0, \dots, \underset{i}{1}, \dots, 0)$, vectors of the canonical basis of $\mathbf{B}^n$

$\mu \boxplus \lambda = \{i | i \in \{1, \dots, n\}, \mu_i \neq \lambda_i\}$

$\Phi^*(\mu) = \overline{\Phi(\overline{\mu})}$, the dual function of Φ

$\Phi^{(k)}(\mu) = \begin{cases} \mu, & \text{if } k = 0, \\ (\underbrace{\Phi \circ \dots \circ \Phi}_{k})(\mu), & \text{if } k \geq 1 \end{cases}$, the $k-$ **iterate** of Φ

$\Phi_i^\lambda(\mu) = \begin{cases} \mu_i, & \text{if } \lambda_i = 0, \\ \Phi_i(\mu), & \text{if } \lambda_i = 1. \end{cases}$, the $\lambda-$ **iterate** of Φ

$\underline{\mu_i}$, we use to underline these coordinates of μ that fulfill $\Phi_i(\mu) \neq \mu_i$

$\overline{\Phi}_\mu = \{i | i \in \{1, \dots, n\}, \mu_i \neq \Phi_i(\mu)\}$

$\mu^- = \{\nu | \nu \in \mathbf{B}^n, \exists \lambda \in \mathbf{B}^n, \Phi^\lambda(\nu) = \mu\}$, the immediate predecessors of μ

$\mu^+ = \{\Phi^\lambda(\mu) | \lambda \in \mathbf{B}^n\}$, the immediate successors of μ

$O^-(\mu) = \{\nu | \nu \in \mathbf{B}^n, \exists \lambda \in \mathbf{B}^n, \dots, \exists \lambda' \in \mathbf{B}^n, (\Phi^\lambda \circ \dots \circ \Phi^{\lambda'})(\nu) = \mu\}$, the predecessors of μ

$O^+(\mu) = \{(\Phi^\lambda \circ \dots \circ \Phi^{\lambda'})(\mu) | \lambda \in \mathbf{B}^n, \dots, \lambda' \in \mathbf{B}^n\}$, the successors of μ

$\theta^\tau(\mu) = \mu \oplus \tau$, the translation with τ

$\Theta_n = \{\theta^\tau | \tau \in \mathbf{B}^n\}$, the set of translations

$[\mu, \lambda] = \{\mu \oplus \underset{i \in A}{\Xi} \varepsilon^i | A \subset \mu \boxplus \lambda\}$, the affine space defined by μ, λ

$[\mu, \lambda) = [\mu, \lambda] \setminus \{\lambda\}$

$(\mu, \lambda] = [\mu, \lambda] \setminus \{\mu\}$

Boolean Functions: Topics in Asynchronicity, First Edition. Serban E. Vlad.

$(\mu, \lambda) = [\mu, \lambda] \backslash \{\mu, \lambda\}$

$Af(\mathbf{B}^n) = \{h | h([\mu, \lambda]) = [h(\mu), h(\lambda)]\}$, the set of functions that are compatible with the affine structure of $\mathbf{B}^n$

$d(\mu, \lambda) = card(\mu \boxplus \lambda)$, the Hamming distance

$(h, h') : \Phi \to \Psi$, the morphism (h, h') is defined, from Φ to Ψ

$Hom(\Phi, \Psi)$, $Iso(\Phi, \Psi)$, $Aut(\Phi)$, the sets of morphisms from Φ to Ψ, of isomorphisms from Φ to Ψ and of automorphisms of Φ

$(g, g') \circ (h, h') = (g \circ h, g' \circ h')$, the composition of the morphisms

$(h, h')^{\frown} : \Phi \to \Psi$, the antimorphism $(h, h')^{\frown}$ is defined, from Φ to Ψ

$Hom^{\frown}(\Phi, \Psi)$, $Iso^{\frown}(\Phi, \Psi)$, $Aut^{\frown}(\Phi)$, the sets of the antimorphisms from Φ to Ψ, of the antiisomorphisms from Φ to Ψ and of the antiautomorphisms of Φ

$(h, h')^{\frown} \circ (f, f') = (h \circ f, h' \circ f')^{\frown}$, $(i, i') \circ (h, h')^{\frown} = (i \circ h, i' \circ h')^{\frown}$, the composition of morphisms with antimorphisms

$(g, g')^{\frown} \circ (h, h')^{\frown} = (g \circ h, g' \circ h')$, the composition of the antimorphisms

tcpo, the technical condition of proper operation

Bibliography

1 Anosov, D.V. and Arnold, V.I. (eds.) (1988). *Dynamical Systems I*, Encyclopedia of Mathematical Sciences, vol. 1. Springer-Verlag.

2 Arrowsmith, D.K. and Place, C.M. (1990). *An Introduction to Dynamical Systems*. Cambridge University Press.

3 Brin, M. and Stuck, G. (2002). *Introduction to Dynamical Systems*. Cambridge University Press.

4 Constantinescu, C.-D. (2003). *Haos, factali si aplicatii*. Pitesti: editura Flower Power (in Romanian).

5 Cortadella, J., Kishinevsky, M., and Kondratyev, A. (1997). Technology mapping of speed-independent circuits based on combinational decomposition and resynthesis. *Proceedings of the European Design and Test Conference ED&TC 97, Paris.*

6 Cortadella, J., Kishinevsky, M., Kondratyev, A. et al. (1997). A region-based theory for state assignment in speed-independent circuits. *IEEE Transactions on Computer-Aided Design of Integrated Circuits and Systems* 16 (8): 793–812.

7 Crama, Y. and Hammer, P.L. (2010). *Boolean Functions: Theory, Algorithms, and Applications*, Encyclopedia of Mathematics and its Applications. Cambridge: Cambridge University Press.

8 Cusick, T.W. and Stanica, P. (2009). *Cryptographic Boolean Functions and Applications*. Academic Press, Elsevier Inc.

9 Danca, M.-F. (2001). *Functia logistica, dinamica, bifurcatie si haos*. Pitesti: editura Universitatii din Pitesti (in Romanian).

10 Devaney, R.L. (1992). *A First Course in Chaotic Dynamical Systems. Theory and Experiment*. Perseus Books Publishing.

11 Easton, R.W. (1998). *Geometric Methods in Discrete Dynamical Systems*. Oxford University Press.

12 Georgescu, A., Moroianu, M., and Oprea, I. (1999). *Teoria Bifurcaţiei, Principii şi Aplicaţii*. Piteşti: Editura Universităţii din Piteşti (in Romanian).

13 Hasselblatt, B. and Katok, A. (2005). *Handbook of Dynamical Systems*, vol. 1. Elsevier.

Boolean Functions: Topics in Asynchronicity, First Edition. Serban E. Vlad.

14 Hirsch, M.W. (2005). *Monotone Dynamical Systems*. https://escholarship.org/uc/item/5wr8t3rq (accessed 06 August 2018).

15 Holmgren, R.A. (1994). *A First Course in Discrete Dynamical Systems*. Springer-Verlag.

16 Ilachinski, A. (2001). *Cellular Automata, A Discrete Universe*. World Scientific.

17 Jost, J. (2005). *Dynamical Systems. Examples of Complex Behaviour*. Springer-Verlag.

18 Kalman, R.E., Falb, P.L., and Arbib, M.A. (1975). *Teoria sistemelor dinamice*. Editura tehnica.

19 Kuznetsov, Yu.A. (1997). *Elements of Applied Bifurcation Theory*, 2e. Springer.

20 Lamb, J.S.W. and Roberts, J.A.G. (1998). Time-reversal symmetry in dynamical systems: a survey. *Physica D* 112: 1–39.

21 Lavagno, L. (1992). Synthesis and testing of bounded wire delay asynchronous circuits from signal transition graphs. PhD thesis. University of California at Berkeley.

22 Lee, J., Adachi, S., Peper, F., and Mashiko, S. (2005). Delay-insensitive computation in asynchronous cellular automata. *Journal of Computer and System Sciences* 70: 201–220.

23 Milnor, J.W. (2006). Attractor. *Scholarpedia* 1 (11): 1815. https://doi.org/10.4249/scholarpedia.1815.

24 Moisil, G.C. (1969). *The Algebraic Theory of Switching Circuits*, 1e English edition. Oxford, New York: Pergamon Press.

25 Mortveit, H. and Reidys, C. (2008). *An Introduction to Sequential Dynamical Systems* (Universitext). Springer.

26 Purdea, I. (1982). *Treatise of Modern Algebra*, vol. II. Bucuresti: Editura Academiei Republicii Socialiste Romania (in Romanian).

27 Seger, C.J. (1991). On the existence of speed-independent circuits. *Theoretical Computer Science* 86: 343–364.

28 Sterpu, M. (2001). *Dinamică şi bifurcaţie pentru două modele van der Pol generalizate*. Piteşti: Editura Universităţii din Pitesti.

29 Trifan, M.P. (2006). *Dinamică şi bifurcaţie în studiul matematic al cancerului*. Piteşti: Editura Pământul.

30 Vlad, S.E. (2007). Boolean dynamical systems. *ROMAI Journal* 3 (2): 277–324.

31 Vlad, S.E. (2009). Universal regular autonomous asynchronous systems: fixed points, equivalencies and dynamical bifurcations. *ROMAI Journal* 5 (1): 131–154.

32 Vlad, S.E. (2010). Universal regular autonomous asynchronous Systems: omega-limit Sets, invariance and Basins of Attraction, Mathematics and its Applications / Annals of AOSR, pp. 249–270.

33 Vlad, S.E. (2011). On the basins of attraction of the regular autonomous asynchronous systems, Acta universitatis apulensis, Mathematics-informatics, special issue. *Proceedings of ICTAMI 2011, Alba Iulia*, pp. 263–286.
34 Vlad, S.E. (2012). *Asynchronous Systems Theory*, 2e. LAP LAMBERT Academic Publishing.
35 Vlad, S.E. (2015). Asynchronous flows: the technical condition of proper operation and its generalization. *International Journal of Computer Research* 22 (4): 435–445.
36 Wiggins, S. (2003). *Introduction to Applied Nonlinear Dynamical Systems and Chaos*, 2e. Springer.
37 Williams, T. (1990). Latency and Throughput Tradeoffs in Self-Timed Speed-Independent Pipelines and Rings. Technical Report: CSL-TR-90-431. Computer Systems Laboratory, Departments of Electrical Engineering and Computer Science, Stanford University.
38 Wolfram, S. (2002). *A New Kind of Science*, 1e. Wolfram Media Inc.

Index